STATISTICS IN SOCIAL STUDIES

LONDON: HUMPHREY MILFORD: OXFORD UNIVERSITY PRESS

STATISTICS IN SOCIAL STUDIES

EDITED BY

STUART A. RICE

Professor of Sociology and Statistics

University of Pennsylvania

FOR THE

Committee on Social Statistics

OF THE

AMERICAN STATISTICAL ASSOCIATION

UNIVERSITY OF PENNSYLVANIA PRESS

PHILADELPHIA

1930

Printed in the United States of America by
Lancaster Press, Inc., Lancaster, Pa.

"In this methodological no-man's land, the statistician must do the best he can. He may have to develop a new technique; in that case he has two chief reliances, first and foremost a sound and wide acquaintance with the field of activity in which he has to operate, and second a good mathematical background, because it is from that that all techniques are developed; but if choice must be made between familiarity with his subject and familiarity with mathematics, I should unhesitatingly prefer the former."

—EDWIN B. WILSON

COMMITTEE ON SOCIAL STATISTICS
AMERICAN STATISTICAL ASSOCIATION

William A. Berridge

F. Stuart Chapin

Neva R. Deardorff

Emil Frankel

Ralph G. Hurlin

Maurice J. Karpf

Philip Klein

Emma A. Winslow

Stuart A. Rice, *Chairman*

FOREWORD

The papers composing this volume have been assembled by the Committee on Social Statistics of the American Statistical Association for a definite purpose. The aim has been to exhibit the nature of the problems that are encountered when the methods of statistics are applied to social and sociological studies. The development of statistical method in the social sciences has been uneven. The time is rapidly approaching when *every* economist, *every* psychologist, and *every* specialist in educational research must at the same time be a statistician. The time is still remote, it would seem, when statistical training will *necessarily* be required of the sociologist, the social worker, the cultural anthropologist and the political scientist. Nevertheless, the trend with respect to investigation in all of these fields is the same.

Differentiation in statistical methodology appears to depend upon the particular characteristics or needs of the subject matter to which it is applied, rather than upon essential dissimilarities of technique. Thus, there are textbooks in business statistics, in educational statistics, and in psychometrics. The concepts and technical methods employed in each are fundamentally similar. But no one has yet ventured to give to a book the title "sociological statistics," "social work statistics," "political statistics," or "anthropological statistics."

This hesitancy is in part attributable, no doubt, to the uncertainty surrounding the nature of the tasks with which each of these subjects, at least the first, is concerned. But it is also due to questions concerning the *applicability* of the fundamental statistical concepts and techniques to these fields. An adaptation of established techniques to new kinds of subject matter is comparatively simple, once it has been clearly shown that the techniques are applicable. It is for this reason that the Committee on Social Statistics has come to regard as its primary function an examination of the problems which arise when such attempts at application are made to social studies. Whether new uniformities of method or approach will ultimately appear therein—such as may have justified a collecting together of various statistical methods and applications under the title *business statistics*—need not be predicted at present. The examination, for the time being at least, must proceed item by item—the "items" consisting of particular problems, or particular sub-divisions of a scientific field, with respect to which some conclusions concerning the possibility and utility of statistical procedure can in each instance be tentatively drawn.

A brief resumé of the Committee's activities to date will provide a background for an understanding of the present publication. The Committee was created as a result of a special gathering, during an annual meeting of the American Statistical Association in New York, December 30, 1925. "About thirty persons interested in the development of a social index" were present.[1] The formal appointment of the Committee, following the recommendation of this gathering, was made early in the following year. It was especially charged with an investigation of "the desirability and feasibility of establishing indexes of dependency and of other variables reflecting social

[1] *Journal of the American Statistical Association*, XXI, 74. (Henceforth abbreviated as *J. A. S. A.*)

conditions." Its continuation was recommended at the next annual meeting with the following definition of purpose: "To act in so far as possible as a clearing house for information about studies concerned with the development and analysis of statistics of social welfare, and to promote and correlate the gathering of representative statistics in the various departments of social welfare activity." [2]

The original concern of the Committee is thus seen to have been with *social welfare* statistics; although, as it notes in its first report to the Association, its name "indicates a broader scope." During its first years of service it shared in the development of several enterprises related to the immediate task. Two of these, under the direction of Ralph G. Hurlin, the Committee's first chairman, are especially worthy of mention. The first is the system of reporting by social agencies which Mr. Hurlin describes in one of the following papers. The second is a study of dependency indexes in the state of Pennsylvania, the results of which are in process of preparation for publication by Hugh S. Carter. At the annual meeting of the Association in December, 1928, a program of papers arranged by the Committee was designed to present a "statistical evaluation of the results of social experiments" in the realm of welfare. The papers presented the following year, at the annual meeting in December, 1929, were similarly concerned with "periodic welfare statistics" and with the question of prediction in social work.

At the beginning of 1929, it was felt by the Committee that the time had arrived for an expansion of its interests, to render them more nearly commensurate with its name. This view was reinforced by accumu-

[2] *J. A. S. A.*, XXII, 105–6.

lating evidence that the American Statistical Association is composed for the most part of persons whose interests are confined to business and economic problems.[3] The Committee felt itself to be in a *liaison* relationship to the Statistical Association, on the one hand, and to such an organization as the American Sociological Society, upon the other. More statisticians should be interested in sociology, or more sociologists should be interested in statistics, or both. It felt that the annual December meetings in 1929 might contribute toward a loosening of the monogamic bond between statistics and economics and the raising of sociology to at least concubinary status. Moreover, the Committee felt the need, implied above, of an appraisal of the extent to which statistical methods have already been developed, utilized, or foreshadowed in a variety of social and sociological studies. It was this question which furnished the *motif* for its program of papers upon otherwise diverse subjects at three joint sessions of the Statistical and Sociological societies at the 1929 meetings. Contributors were urged to direct their discussions to questions related to *statistical method*, rather than to the content or detailed results of the studies considered.

All of the papers which follow except the first were planned for the December meetings, but a portion only were actually upon the program. These have been edited with the intent of removing all transient elements referring to the time and place of oral presentation. The thread of unity binding the series together is to be found in the purpose of the Committee rather than in the organization of topics

[3] *J. A. S. A.*, XXIV, 303–6; *ibid.*, XXV, June, 1930, note by Stuart A. Rice and Morris Green.

discussed. There is no claim of logical sequence, or of logical division of a universe of discourse, to be found within the collection of topics themselves. It is for this reason that the Committee and the Editor feel it to be unwise to attempt any hasty generalizations from the several contributions here assembled. They are presented to the reader with a minimum of editorial revision or comment.

A special word of explanation is required with respect to the three papers which deal with the subject of prohibition. Some hesitation was felt by the Committee in including this topic upon its program. The subject itself is highly controversial. The interests of the Committee, on the other hand, were wholly non-controversial, since they concerned *methods of research* rather than the pros and cons of opinion with respect to the subject. At the same time, it was felt that any mention of the subject would tend to be interpreted by interested persons in a partisan manner, unless the Committee took considerable care to safeguard its neutral attitude with respect to the merits of the prohibition issue. Moreover, it seemed difficult to find a competent statistician, familiar with research in the field, who was at the same time in a position of neutrality before the public with respect to it. Nor did the Committee wish to exhibit timidity in the case of a field of study so important, by ejecting the topic from its purview.

As a solution to the dilemma, it was decided to procure methodological papers from competent social scientists and statisticians, who would at the same time be recognized as *partisans on opposite sides* of the wet and dry controversy itself. It was hoped in this manner to "cancel out" the factor of partisanship. The success of this aspiration must be appraised by the reader. Since the first of the three papers on the subject, that by Gebhart, represents a specific subject of criticism and attack in the remaining two, by Fisher and Feldman, no attempt has been made to eliminate the controversial aspect which connects them. Each of the other papers in the book stands alone as a unit. These three stand together as a group.

The Editor's acknowledgments are here given to those who have had a share in the plans and preparations of the present volume and the program from which it originated. These include his fellow members upon the Committee itself, and in particular his predecessor as chairman, Dr. Ralph G. Hurlin. All of the contributors have the Editor's gratitude for the cordiality of their coöperation, particularly in the forfeiture of rights to royalties, which are conveyed to the Committee. Finally, the officers of the American Statistical Association and the American Sociological Society have rendered essential aid by giving permission for the publication in this form of papers delivered under their auspices, and by assisting in bringing the volume to the attention of their readers.

STUART A. RICE

CONTENTS

I

THE HISTORICO-STATISTICAL APPROACH TO SOCIAL STUDIES

STUART A. RICE

SOCIAL change, ever more rapid, and involving continual readjustment by individuals and groups, is receiving increasing attention from social scientists. They are being called upon to find ways of determining regularities of time sequence in social phenomena as a basis for predicting what the future movements will be. Men have long recognized that the present could seldom be understood without a knowledge of the past. This recognition has resulted in the *genetic* and *historical* approaches to the study of contemporary social phenomena. But these approaches have usually taken the form of *description in general terms* of past societal states; and of historical *narrative*, also in general terms, of the changes ensuing up to the present. So far as these approaches have led to prediction, the latter has been highly speculative.

General description and general narrative fall short, in several respects, of meeting the present demand for dependable analyses of social movements. In the first place, they lack precision. Secondly, since they cannot employ precise criteria of relevancy, they tend to select and emphasize particular elements and events which may be striking to the attention, but which are relatively unimportant from the standpoint of later interests. Present-day social scientists would exchange bushels of ancient annals of kings and bishops for a few more authenticated facts concerning such plebeian matters as roads, interest rates, and peasant customs. Thirdly, and to much the same effect, description and narrative tend to deal with individuals rather than with mass situations. The latter may escape recording by virtue of the very fact of universality. Lastly, important relationships may be concealed from casual view, to be disclosed only when the methods of statistical analysis have been developed and applied.

To the historical approach, therefore, has in recent times been conjoined the statistical—a conjunction which seems to justify the title of the present discussion. It is in the analysis of historical or time series of data [1] that the outstanding statistical achievements of recent years have been made, and with respect to which, perhaps, the most engrossing problems of statistical methodology remain to be solved. Especially by throwing new light upon social relationships has the statistical approach given added importance to the historical, and the underlying points of view supplied by each have resulted in mutually effective supplementation.[2]

[1] The term "series" as used here is defined by Chaddock in his *Principles and Methods of Statistics* as "a *general term* used to describe a succession of enumerations, measures, or estimates without specifying the particular form of arrangement of the data." Chaddock also notes that "types of series should be distinguished according to the fundamental relations of the items in time and space."

[2] The historian will continue, no doubt, to retain his interest in *unique events*, while the statistician will employ such events only as units in numerical summarization. It is with respect to underlying assumptions and concepts concerning social change that each has contributed to the thought of the other. The notion that the present is to be understood as the result of orderly deterministic development out of the past seems to me to be fairly termed an *historical* idea, though some might prefer the term *genetical*, which for the sake of convenience has been dropped from further use in the present discussion. The

It is particularly within the fields of economics and business administration that these developments of method have occurred. The principal aim of the business statistician is to separate the combined effects of various causative influences which lie behind historical series of data; thereby to understand these influences better; and to use the results of his analysis for prediction and for control of the future course of events.

The considerations which have led to the present emphasis upon historical statistics in the economic and business world seem to apply with equal force to a study of social phenomena of numerous other kinds. Cultural and institutional tendencies which may be currently evaluated as "progressive" or "retrogressive," relationships existing among particular trends of social habit or custom, and especially relationships between specific social changes and more general movements in the social and economic life of the people, can be ascertained best by a combination of the historical and the statistical approaches. Knowledge of such tendencies and relationships provides a foundation for the power of prediction and social control, the securing of which may be regarded as the ultimate objective of scientific social research.

The causal influences to which statisticians are accustomed to look for an explanation of alterations in business are usually placed in four classes. These classes may be termed *secular trend*, *cyclicality*, *seasonality*, and *chance disturbance*. *Trend* is conceived as the result of constant or cumulative influences making for change and operating with steadiness over a relatively long period of time. There was, for ex-

statistician, on the other hand, seems to have presented the historian with a new notion when he contends that the major aspects of this orderly change may, and should, be expressed in quantitative terms.

ample, a gradual growth for many years, up to the middle of the decade just closed, in the American family consumption of milk.[3] Part of this growth resulted from increase of population. When this factor was eliminated by reducing the consumption to a per capita basis, there was still a growth. This was doubtless to be explained by slowly changing popular habits with respect to food and drink, induced in part by the propaganda of health organizations. Perhaps it was induced also, in part, by rising family incomes.

However, this upward trend did not prevent periods of several years each in which the per capita consumption of milk appeared to be declining, or in which the *rate of increase* in consumption appeared to be declining. In other words, superimposed upon the long-time, slowly changing influences was a second set of factors of less enduring character which induced *cycles* about the line of trend.[4] In recent years the concept of a *business cycle* has become familiar to the American public. What is not so clearly understood is that the cycle itself is measured from a trend. The trend may be up or down, and it may be changing arithmetically or geometrically, but at any particular time it is regarded as the normal situation. Thus, at the lowest ebb of a cyclical depression in the milk industry, the per capita sales were nevertheless larger than in the midst of a period of prosperity a few years preceding.

[3] I.e., excluding milk products. More recent data on milk consumption have not been analyzed by the writer. It seems probable that the upward trend has continued.

[4] We employ the usual terminology, which implies a system of graphic coördinates. Time is represented as moving from left to right; quantities or frequencies, as increasing upward from a zero base line. The line of trend would represent the "best fit" to points indicating quantities at given intervals of time.

The point is that the *normal* situation had itself changed in an upward direction.[5]

Just as influences which are cyclical in character may superimpose their effects upon those which result from the long-time causes, so also a third group of influences related to the *season* may further superimpose effects upon the other two. An annual rhythm in the ice cream business, due to seasonal changes in demand, offers a familiar illustration.

The fourth type of influence, the results of which may appear in a business series, consists of those *fortuitous* occurrences of special or isolated character which do not appear to operate or recur in any fixed order. They represent the type of *unique events* to which an historian is inclined to devote particular attention. Their effects appear in a time series as a residue, when the constant or recurrent types of influence have been eliminated. Among such disturbances are floods, wars, and disasters generally. In addition, there are lesser causes, the results of individual chance or fortune, the effects of which cannot be so easily identified.

It seems likely from *a priori* considerations that the four types of factors identified in the data of business [6] will be encountered also in the data of social movements other than business. Ogburn, Thomas, Hexter, and others,[7] have found a close relationship

[5] Since this was written, Kuznets has distinguished between *primary* secular variations and *secondary* secular variations in economic data, the latter being intermediate in duration of period between the former and the business cycle. Cf. Kuznets, Simon S., *Secular Movements in Production and Prices.* Boston: Houghton Mifflin Co., 1930.

[6] *Five* types of factors, if the discrimination by Kuznets referred to in footnote 5 be included.

[7] Ogburn, William F., and Thomas, Dorothy, "The Influence of the Business Cycle on Certain Social Conditions," *J. A. S. A.*, 1922, pp. 324–40. Thomas, Dorothy Swaine, *Social Aspects of the Business*

between cycles of business activity and such social phenomena as marriage, divorce, births, deaths, still births, suicides, church attendance, and crime.

There is no sufficient reason to suppose that types of recurrent influence other than those already recognized may not also be found in social data. Kroeber has discovered a long rhythm with respect to fashions of dress;[8] while Chapin, on the basis of considerable evidence, has formed a theory of (long-period) "synchronous culture cycles."[9] At the New York Municipal Lodging House, recurrent sub-seasonal (monthly and weekly) fluctuations in patronage were found to exist.[10] These were related, to some extent, to municipal and departmental regulations; but they were also related to such an established folkway as the Saturday night bath. Moreover, they were of no less practical importance than the longer movements. In analyzing data of annual political elections for members of the New Jersey legislature, the writer found a characteristic cycle of "party turnover," apparently related to the business cycle; but he was also compelled to take account of a recurring quadrennial influence (in addition to trend and cycle) which had to do with the presidential election. "Correction" for this was made by an "index of quadrennial variation."[11] It would be a mistake in an untried

Cycle. New York: Alfred A. Knopf, 1927. Hexter, Maurice Beck, *Social Consequences of Business Cycles.* Boston: Houghton Mifflin Co., 1925.

[8] Kroeber, A. L., "On the Principles of Order in Civilization as Exemplified by Changes in Fashion," *American Anthropologist*, N. S., 21, No. 3, July–September, 1919.

[9] Chapin, F. Stuart, *Cultural Change.* New York: The Century Co., 1928.

[10] During the writer's superintendency of that institution, 1916–1918.

[11] Rice, Stuart A., *Quantitative Methods in Politics.* New York: Alfred A. Knopf, 1928, Part VI. In particular, Chapter XX.

field to accept uncritically the assumptions of the business statistician as to the types of causative influence which may be combined in the data under examination.

The uses to which historical statistics of social and sociological data may be put are similar to the uses found for business and economic statistics. The difference is largely one of incentive. The business statistician or his employer usually has pecuniary profit, or avoidance of loss, as a specific objective. Moreover, he is likely to have in view a particular business situation. Both of these factors provide a stimulus for his work and give point to his findings. The "social" statistician may have the much less specific objective of "serving the public good" in some general or indirect manner; and the public is too often an unappreciative employer. There are outstanding examples of analyses of sociological time series for general public ends, such as the population forecasts of Raymond Pearl, Louis Dublin, and others; but even these are not far removed from particular economic interests. The possible utility of statistical analysis in connection with public policy, and in connection with generalized programs of social well-being, is still to be "sold" to most of the public officials and administrators of social activities outside of the economic sphere.

A single illustration of this potential utility, related to public welfare administration, will be given here because it fell under the writer's immediate observation. Between the years 1914 and 1918 inclusive, the Municipal Lodging House of New York, as already noted, exhibited several types of regular fluctuations in the daily number of inmates. These were weekly, monthly, seasonal, and cyclical in character. They were apparently related to social folkways, to the weather, to administrative regulations, and to business fluctuations, particularly in certain industries and occupations to which homeless men are accustomed to look for employment. Further, there seems to have been some suggestion of a trend, although this would have been affected sharply by changes in legislative or administrative policy. Moreover, a trend would have been obscured, during the period mentioned, by fortuitous disturbances due to the war. If a trend upward existed, it may or may not have been due, as critics contended, to a spreading knowledge in other communities of the shelter offered by New York City to homeless men. The aggregate effect of all these influences, it should be emphasized, appeared in a single daily total of admissions. There was no statistical analysis by accredited methods of the series formed by these totals.

In the winter of 1913–1914 the city of New York was confronted with a severe unemployment crisis. The Municipal Lodging House lacked capacity to house the throngs of homeless men and women who sought its shelter. An annex was hastily prepared by enclosing the East 23rd Street recreation pier, and installing beds, blankets, stoves for heating, and other accessory equipment. In this manner, approximately fifteen hundred additional accommodations were provided, above the normal capacity of the Municipal Lodging House itself, which was in excess of eight hundred. With the arrival of spring, the pier was dismantled and made ready for its normal functions. It was again put to use during the following winter (1914–1915) and again dismantled.

With the approach of the third winter (1915–1916) the question arose as to whether or not the pier would again be required. As cold weather came on, the daily admissions steadily mounted. Late in December, the conjunction of a cold night, the normal Saturday peak, and the normal seasonal growth, sent the number of admissions to a high mark, but still within the capacity of the Municipal Lodging House itself. On the assumption that another period of stress was imminent, steps were immediately taken to fit up once more the annex on the pier. Yet the latter, laboriously and expensively prepared, was not used. Those in command had failed, even in their own minds, to separate the aggregate "census" into the effects of the separate influences which were involved. While the seasonal and weekly tendencies were "up," the cyclical movement was distinctly "down"; but this was not known at the time, for no attempt had yet been made to graph the series of data which would have established the fact. It subsequently became apparent that *if* statistical methods had been used, even without refined analysis, the expensive preparations for a burden that failed to arrive would not rationally have been undertaken.

While the foregoing suggests that the Municipal Lodging House data themselves might have provided a basis for prediction concerning the needs of that institution, the establishment of relationships between these and other data of social and economic character might also have facilitated prediction.

It cannot be denied that in comparison with business and economic statistics the development of social and sociological statistics is handicapped in numerous directions. Uniform reporting systems, with central collection agencies and agreed-upon definitions of units, will have to be developed in some instances from the ground up. Illustrating such developments are various endeavors to procure uniformity of definition of crimes and systematic reporting of criminal statistics; also, the noteworthy attempt to collect statistics of social case working agencies undertaken by the Russell Sage Foundation. Reference to one or another of these endeavors is made in three of the following papers.

Another handicap upon the development of social statistics concerns the reporting period. Monthly data are essential to refined analysis. A knowledge of seasonal factors may in itself have practical importance; moreover, turning points in the cycles may be obscured when annual figures only, or even quarterly figures, are available. But, while monthly reporting has become familiar in business, it is not so familiar in respect to numerous other social data. Furthermore, the endeavor to procure monthly data would in some cases be quite impracticable and would defeat its ends. An illustration of this would be found in the field of social attitudes and public opinion, discussed in another of the following papers. Any effort to obtain monthly indexes of attitudinal states, if successful, would be certain to change these states in the process. The political statistician is fortunate to find recurrent and comparable data collected annually.[12]

Other problems involve the combination of different series into index numbers, although here the perplexities are not unlike those encountered in connection with the more or less standardized concepts and techniques of business statistics. An illustration will be considered in some detail.

[12] Cf. Rice, *op. cit.*, Part VI. Distinct curvilinear and linear trends and cycles were found in the median age on election, and in the median length of service, of American congressmen, on the basis of *biennial* data.

Suppose it is desired to establish a general index of social welfare, on the basis of data regularly reported by public and private social agencies.[13] It is obvious that unless two series tend to fluctuate, to some degree, in direct conformity with each other, the result of their inclusion together in an index will be that each tends to cancel out the effects of the other. In some cases this will be desired. For example, if the number of placements of children in institutions was found to bear an inverse relationship to the number of placements in family homes, the inclusion of both series in an index of dependency would nevertheless be desired, for the relationship of the two is supplementary in character. Suppose, on the other hand, that a negative correlation, or no correlation, was found between first admissions to mental hospitals and the proportion of free days in general hospitals. This would seem to indicate that the forces governing the fluctuations in each were of unrelated or different character. It would be improper, therefore, to include both in an index which was designed to reflect the influence of a single type of force.

It must be kept in mind that an index is a *representative* figure. It stands for some quantity, degree of activity, value, or situation, which cannot itself be expressed directly. In a fundamental or philosophical sense, all measurements are indexes of a hypothetical world of reality beyond our reach. But in a more usual or practical sense, index numbers are required, first, when the thing in which we are interested is intangible or poorly defined (for example, individual intelligence), and second, when we are attempting to make a net or total statement concerning a complex

[13] Cf. the discussion of dependency indexes by Dr. Hurlin in Chapter IV.

group of tangible things or events (for example, the cost of living, which is ultimately not only a matter of standards but of individual purchases and sales). In the case of intelligence, we use the "Intelligence Quotient" or "I.Q." as an index based upon the organism's behavior under controlled conditions. To determine the "cost of living," we build up an index which is a form of average based upon a sampling of business transactions.

Thus, there seem to be two types of index numbers, the term being usually reserved for the second—that which pertains to a group of related phenomena rather than to intangibilities within the individual case. The related phenomena which we desire in some manner to summate or average are necessarily regarded as belonging to a common field. Now it is important to recognize that this field is delimited according to the interest or purpose which we have in view. A cost of living index designed for city dwellers will be based upon transactions in a different variety of commodities or services than those which are included in an index for farm dwellers. Even if, in the composition of two index numbers of prices, the commodities should be the same, the relative importance (or "weighting") assigned to each commodity in the averaging process may differ according to the respective ends in view. In other words, there is no natural "field" of phenomena (in the sense above used) apart from the purpose in the mind of the person segregating it from other "fields."

In the present illustration, the general field with which we are presumably concerned includes both social work and the social problems with which social work is engaged. These are logically dissimilar and even, perhaps, antithetical. Social problems (dependency, delinquency, etc.) might be compared to the water in a pond. Social work might then correspond to efforts expended to drain the water out. The amount of the efforts at a particular time would bear no *necessary* relationship to the amount of water in the pond. However, the situation might be such that increased volume of water in the pond increased the drainage. The latter would then provide an index of *changes* in the former, though not of the former itself. On the other hand, increased drainage might *reduce* the water level, in which case we might still have an index, but it would be negatively related to changes in pond volume.

This analogy suggests some of the situations which might be encountered in setting up social index numbers. A great deal depends upon the particular nature of our interests. The data provided by a public department, charged by law with relief of necessitous persons, are likely to prove more useful in gauging changes in the fundamental *problems* which social work attacks than will records from private social agencies. But these same public records will be of little or no value in constructing indexes of the *amount of effort* expended on these problems, i.e., of social work itself.

The problems which social work attacks are a conglomerate of many different things, the common element among which has never been clearly perceived. The question arises whether, if we attempt to establish indexes, we are really dealing with one field or several. There has been discussion of disease epidemics, of "hard times," of "crime waves," and of "waves of insanity." All of these expressions imply the existence of definite modes of variation in certain realms of social phenomena. In setting up a social index we have to decide whether crime, dependency, sickness, and mental abnormality in its various forms, can all be regarded as aspects of the same field; and hence, whether all may contribute to the construction of a common index, or whether they should be kept in separate compartments of a statistical program. The answer is found in a reiteration of the principle that the purposes sought by the establishment of the index or indexes must control. It is therefore necessary to examine these possible purposes.

It seems to the writer that two aims might be in the minds of those who would be concerned in establishing such an index. These may be referred to as the "bookkeeping" and the "curative" interest, respectively. Social workers are agreed in desiring a better world. Even if they would not wholly concur in its specifications, they are in general concurrence as to the evils in our present society which require eradication. Dependency, sickness, crime and mental abnormality, are among these evils. Without respect to their causation, all are to be destroyed if possible. Part of the social worker's task is to arouse the public to the menace of these evils, and to persuade it to support the various efforts for their elimination, whether by cure or prevention. Consider a simple analogy: A man of many widely divergent business interests is seeking to decrease his liabilities. Some of his ventures are gaining while others are losing, each from a wholly different cause. An index of his economic security, and hence a general guide to action, is provided by the final balance upon his books. A similar interest in the progress of the campaign waged upon social ill-being on all fronts would justify the inclusion of the traditional "three

D's" (dependency, delinquency, defectiveness) in a single *index of social atypicality.*

A second, and perhaps a stronger motive for the establishment of social indexes, however, relates to the hope for *ultimate control* over the social problems involved. An index is a tool which may be used in the effort to attain such control, provided it has been constructed with reference to the underlying forces which produce the problems. For this purpose, there must be homogeneity among the latter. Otherwise, changes in the index will bear no clear relationship to changes in the causal influences at work; the index will not disclose which of related groups of causes may have been altered. Thus, the business man will not seek to relate the changes in his own enterprise to the business cycle in general, but rather to some series of business indexes which presumably are determined by the same set of influences which affect his own affairs.

Again, there will be two general types of interest which would govern the selection of particular series for the indexes derived. First, there would be an interest in changes with respect to *intake*—*new* dependency, *new* delinquency, etc. Second, there would be an interest in the cumulative effect—the present *load* of dependency or delinquency. These interests would roughly correspond to the economic interests in indexes of production and in indexes of stocks on hand.

A final problem concerning social index numbers which will receive attention here concerns the question of weighting. It has become well understood that weighting in some manner is unavoidable: if all series be regarded as of equal importance, this in itself constitutes a weighting process.

The writer is impressed, however, with an argument advanced by Hexter:[14] The series with which we deal are in few cases more than *samples* of situations, the full extent, or the comparative seriousness of which, are undisclosed. To present an illustration from another field, suppose that we are interested in the amount of dust carried in the air on the streets of Philadelphia. Samples of air are procured at various locations. Some of these samples may be of larger volume than others, but we are not thereby justified in giving greater weight to the former in any index of dust content established for the city as a whole. In fact, a determination of the total dust content of the air in the city will not be secured from any such index. It will not be difficult, however, to determine *variations* in that content.

Another analogy is to be found with respect to indexes of unemployment. Hurlin and Berridge reached the conclusion that the *amount* of unemployment was unascertainable, as a practicable measure.[15] *Variations* in its amount, on the other hand, are not difficult to discover.

Neither analogy is entirely applicable to the problem of weighting the several series contained in a social index. In the case of the second, especially, a weighting system would probably be employed. Nevertheless, the analogies hold in the respect that we cannot easily measure the total *amount* of dependency, delinquency, or morbidity, but only *changes* in these. Nor can we say that, because a vagrant is sheltered for a single night in an almshouse, while a pauper remains there throughout the year, hence the pauper is

[14] In a conference of the Committee on Social Statistics.

[15] Hurlin, Ralph G. and Berridge, William A., *Employment Statistics for the United States*. Their statement was made prior to the development of plans for a census enumeration of the unemployed.

365 times more important than the vagrant as an indicator of dependency. By avoiding one type of pitfall and deciding to weight the two equally we, of course, fall into another: The vagrant may be counted *several* times while the pauper is counted but once. Nevertheless, any method of assigning more weight to one than to the other must be based upon the premise that the problem of pauperism is *more important* (in some definite ratio of importance) than the problem of vagrancy. In the absence of some ground for such an assumption, it seems safer in most cases to allow equal weights to the series included.

There are other special problems in the establishment of a social index which would not be encountered in the development of indexes of business. The question arises, for example, whether the work done by social agencies, either public or private, is a function (*a*) of the magnitude of the social maladjustment itself, (*b*) of the social interest in the maladjustment, or (*c*) of the provision made for meeting the situation caused by the maladjustment. Those who have urged the desirability of general indexes of social welfare seem at times to take for granted the first of these possible functions. The validity of the assumption, however, would have to be defended on logical grounds in any given instance. As to the second, no one would suppose, for example, that there was no need for playgrounds in an urban community prior to the time when facilities were provided, or prior to the time when organizations were formed to promote recreation. As to the third possibility, it is relevant to note that family societies would often accept more cases if they were granted more funds with which to work. On the whole, it appears that series concerning public and private social work and poor relief are more nearly representative of *interest* and *provision made* than of *fundamental need*.

How, then, can series be obtained which will indicate the need? Applications for assistance will not suffice, for if it becomes known that cases are being turned away applications may not be received which would otherwise be made. On the other hand, a change in the attitude of an agency (a "warmer" or "friendlier" attitude, for example) may "bring in" applications.

Widespread changes in social organization, and in the resulting attitudes toward personal responsibility, have tended to shift the burden of many social needs. For example, a state of dependency which in a former period might have been handled without public attention by relatives or neighbors may now be made a responsibility of a social agency. Does this shift of burden mean merely a shift, or does the shift itself constitute part of the increased problem? If the work of the agency be regarded as an index of the problem, then before there was an agency there was no problem. *Reductio ad absurdem.* When a Municipal Lodging House is established, the number of its applicants is included in the index of dependency. But what of the community that has no Municipal Lodging House? I send the panhandler there if I know such an agency exists; otherwise, I may give him a quarter myself.

Another set of questions which would have to be answered in the course of establishing an index of social well-being would relate to the type of recurrent movements involved in the series employed. So far as the problems with which social work has to deal are concerned, we should expect, *a priori*, that dependency would conform, in an inverse relationship, to the general movements of economic prosperity.

The expectation is clear in the case of the cyclical changes. Moreover, we would expect that a general long-term improvement in business prosperity would have resulted in a corresponding decline in those forms of social ill-being which are thought of as dependency.

In the case of delinquency, a general relationship, either positive or negative, to either business changes or dependency, is problematical. The question is chronically in a stage of public controversy. The new generation is always "going to the bad" (trend) in the opinion of its elders. Easy wealth is demoralizing to conduct, in the opinion of moralists (cycle), while poverty is conducive to crime (cycle) according to a more socialistic trend of thought. Delinquency will doubtless have to be broken up into its parts before exact statements can be made concerning it. Some crimes may increase and others diminish with the business cycle.

As to morbidity, there is much more occasion for optimism as to a downward trend. But here again the data are likely to be misleading. The real improvement represented by better reporting or increased hospitalization may boost the apparent trend. As to the cycle, previous work has left its nature an open question. The statistics of social work as an activity or activities will not involve so many difficulties, provided the interest is in the work itself and not in the problems attacked. That is, it will be possible to state that our efforts have slackened or increased in a definite manner. But will this give us any clue as to what it is all about? As to the modes of fluctuation, anything from a high positive to a high negative correlation with the business cycle or with dependency indexes is possible, according to the particular series of social work data that is examined. It is possible that states of mind which register themselves in political opinion and in generous interest in the well being of others are relatively independent of prosperity and depression, and are dependent upon factors as yet unknown or little understood.

The possible construction of a general index of social well-being has been discussed at length, because it seemed that some of the unique questions which would be encountered in the application of the methods of historical statistics to social studies would appear more sharply in the light of a concrete proposal. Other applications to other types of social studies would develop equally unique problems. Some of these problems would be inherent in the nature of the material that it was sought to organize. This would be true, for example, of social attitudes and public opinion. In other cases the difficulties would adhere primarily to inadequacies of the reporting mechanism. This would probably be true in the case of statistics of crime, where deficiencies of this kind are already being remedied. But in any field of social study, progress in the development of the historico-statistical approach will probably come about as a result of more earnest attempts to utilize series that may now be available. In the process of utilization, the shortcomings of the data will become more apparent; but so also will the possibilities of strengthening the data by improvements in the methods of their collection.

II

STATISTICAL STUDIES OF MARRIAGE AND THE FAMILY

WILLIAM F. OGBURN

PERHAPS the most important statistical need in studies of the family and marriage is, first, an improvement in accuracy; and second, the collection of data on aspects of the family upon which statistics are not now available. Marriage statistics are generally accurate and reliable. Exceptions occur in the case of Negroes, and in the case of divorce data, particularly in that rather large number of states which have no state registration law for divorce. Statistics on the family, as now tabulated, are quite incomplete and inadequate. Our decennial census should furnish us with data on the sizes of families classified in several new ways: according to ages of husbands and wives; according to ages at marriage; by urban and rural regions; by nativity groups; and also, in certain cases, by occupations or occupation classes. The total number of births and deaths should also be known for the study of fertility rates. There should be, moreover, statistics of the composition of households and statistics of broken homes.

The most obvious statistical methods to be employed in dealing with such data are, of course, those of classification in adequately defined categories. The statistics of the different countries are not as fully comparable as they would be if common agreement could be reached among the nations as to the categories of classification to be used by each. In addition, there are two methods, in particular, which I should like to consider in the remainder of this paper.

One of these presents itself when we wish to compare the percentages of the population married. For instance, in 1890, 55.3 percent of the population fifteen years of age and over was married; in 1920, 59.9 percent was married. There was thus an increase in the population married of 4.6 percent. But one wonders whether such an increase mirrors a change in our social life or whether it may not be due to the fact that there were relatively fewer younger people in the population of 1920 than in 1890. For smaller proportions of our young people are married as compared with our older people. An examination of the age distribution of the population in 1890 and in 1920 shows this to be the case. In the latter year, there are larger proportions of the population over thirty-five years old and smaller proportions under thirty-five years. Hence, the larger percentage married in 1920 may be due not to changing social habits, but rather to the fact that the population is older. How can this factor be measured?

The answer is suggested by a practice used for eliminating the influence of the age factor in comparing general death rates; namely, employment of the device known as the standard population. This seems applicable to marriage rates as well. Thus, if the population of 1920 is distributed in each age group in the same proportions as existed in 1890, age is thus eliminated as a variable factor. If we then apply the specific marriage rates in each age group in 1920 to the new age distribution, and compute the totals, we get the percentage of the 1920 population that would have been married if the age distribution had been the same in 1920 as in 1890. When this is done, the percentage married is found to be 56.8 percent instead of the 59.9. In other words, 3.1 percent out of the 4.6 percent increase in percentage married is due to the fact that the population has grown older.

A few comments might be made on this procedure. When the age groups are too wide, particularly in the lower periods, the device is not so satisfactory. Another point seems to me to be this. The device of a standard age distribution is applicable only when there is no functional relationship between the specific rates and the relative frequencies of cases in the groups from which the specific rates are determined. For instance, one could not eliminate the influence of the sex ratio factor, in the same manner as the influence of the age factor, by using a standardized sex ratio in 1890 and in 1920. This follows because the specific marriage rates for each sex are dependent on the sex ratio. But this objection does not apply to a marriage rate or to a death rate for a particular age period. The specific rates do not depend on the ratio between the proportions in one age group and those of another. For instance, the correlation between the specific marriage rates by cities for the age period fifteen to twenty-four and the proportions of the population which were fifteen to twenty-four years old was + .08 —a coefficient without significance.

In a similar way, it has been possible to show that if in 1920 the race and nativity distribution of the population had been the same as in 1890, then there would have been 60.4 percent married instead of 59.9. The influence of race changes was, then, to diminish marriages. So, also, it was possible to measure the influence on the percentages married of the factor of urban growth from 1890 to 1920. It was believed that this factor had been important, but its influence was found to have been very small. For if the same urban-rural distribution had existed in 1920 as in 1890, then the proportions married would have been 60.0 instead of 59.9.

Similar comparisons can be made between cities and states. But no satisfactory method seems apparent whereby the three factors of age distribution, race and nativity distribution, and urban growth, can be eliminated at one time. Naturally, it would be desirable to measure the combined influence of all three factors on the percentages married.

A slight variation of this device was used in measuring the relationship between marriage and insanity. In this case we have the age distribution of the insane, but not the specific marriage rates of the insane by age groups. But we do have the specific marriage rates by age groups, together with the age distribution, for the general population. We can thus give to the general population the age distribution of the insane, instead of following the more readily suggested process of giving the insane population the age distribution of the general population. The specific rates for marital status can then be applied to the new age distribution of the general population, and thus the age distribution is eliminated. The result shows a higher percentage of the unmarried among the insane when the age factor is held constant.

The second technique which I wish to consider in connection with marriage and the family is that of partial correlation. The problem here is not different from that of the use of partial correlation in general, except, perhaps, in two respects. One of these turns on the unit used in the correlation tables. The possible units are cities, counties, or states. The difficulty of using a geographical unit in correlation lies in the fact that there are often so many other possible factors which ought to be taken into consideration. The larger the unit, the more this is true. The state is a questionable unit because of its great size and the many possible factors that may be identified with it. A very good illustration is where one of the variables taken into account is the sex ratio. The sex ratio may very well be quite different in the smaller local units than in the state as a whole; and it is probably the sex ratio of a local unit that affects marriage more than the sex ratio of a whole state. Many investigators have used states as units, but I question whether much confidence can be put in the results from such correlations. Furthermore, when states are classified by social data, they seldom appear in even approximate normal-probability distribution. Still another objection is that a sample of only forty-eight is too small.

Despite these criticisms of correlation methods with marriage data when the locality is used as a unit, there are certain evidences that true indicators of relationship are provided by correlation coefficients when the localities are cities. For instance, we find the simple coefficient of correlation, with cities as the unit, between the percentage of women married and the percentage of women employed to be − .66, which corresponds very well with the actual census figures showing that only 23 percent of employed females are married.

Another illustration is pertinent. Suppose we wanted to correlate, for 170 cities scattered all over the United States, the percentage of divorced persons and the percentage of the population which is foreign born. The objection is at once made that we cannot do this since the different states have different divorce laws. South Carolina permits no divorce, and in Nevada it is obtained rather easily. This is true. But we may say that the nature of the divorce statutes under which the different cities operate is simply a very important third variable that ought to be held constant. Since we have no measure for it, we may try a simple correlation between percent divorced and percent foreign born, without holding this factor of divorce laws constant, and see what we get. The result of the correlation between the percent of foreign born and the percent of divorced persons for 170 cities is − 0.45. This corresponds with the actual census data, which show that the number of divorced persons reported among the foreign born is from one-third to one-half of that which is reported for the rest of the population. Similarly, we find that the percentage of divorced persons in these cities is correlated with the birth rate to the extent of − .48. We know, of course, that divorce is much more frequent among childless couples.

Another serious difficulty, somewhat peculiar to marriage and family data, in the use of correlation technique is the difficulty of getting a single variable which will represent the age distribution. This difficulty is quite serious because of the great influence which age has on marriage and the great variability of age distribution from place to place. It is possible, of course, to make correlations with specific age groups. In some cases, I have used the percentage of the population fifteen or more years old who are fifteen to nineteen or fifteen to twenty-four years. On another occasion, I have used the percentage of the male population fifteen years old and over who are thirty-five to sixty-four and of the female population who are twenty-five to fifty-four, these being the ages of maximum marriage for the two sexes respectively. It would, of course, be desirable to get a more refined measure of the age distribution.

An illustration of the importance of the age factor is in the correlation of the birth rate and the proportion of young women (fifteen to twenty-four years old) who are married. The correlation is − .14. Evidently the age distribution of married women is quite a factor in the birth rate, particularly since the probability of a birth is greater with younger women. So when the age factor among married women—here taken as the percentage of married women who are fifteen to twenty-four years old among married women who are fifteen to forty-four years old—is held constant, the correlation is raised from − .14 to − .45. So the age factor is often an important variable to hold constant.

This illustration points to a need of interpretation which is present whenever correlation is employed, but which is particularly required with correlations among marriage data when localities provide the units. After all, correlation tables are only a framework enabling one to see the data better—and data are always symbols that stand for something else. When it is found that the birth rate becomes lower as the proportions of young females married become higher, what does this relationship mean and why should it exist? It would seem as though the proportion of young women married would have an influence on the *number of births*, but not on the *birth rate*. On the other hand, could the birth rate have an influence on the percentage of young women married? Since births are generally events subsequent to marriage, one would think they could not influence marriage, any more than statures of sons could influence statures of fathers. But if a low birth rate symbolized an extensive use of contraceptives, then the causal connection would be thinkable. This, however, is an inference. To prove it would require statistics of the use of contraceptives, of which there are none.

Another illustration of the speculative nature of the meaning of partial correlation coefficients is the correlation between the percentage of women employed and the percentage of women married, as a possible indicator of the influence of the employment of women on marriage. The correlation, as we have previously said, is – .66; but this simple correlation coefficient shows no influence of employment of women on marriage. If the employment of women encouraged marriage there could still be a negative correlation of – .66 between the percentage of women employed and the percentage of women married, on the conditions that the women who married left their jobs and that these jobs were filled by single women. In this case we should have, in all probability, an influx of single women to the cities involved, a condition which might be indicated by the sex ratio. (The sex ratio might also indicate an exodus of males, which would have the same effect as an influx of females.)

When the sex ratio is held constant, the correlation between women employed and percent married drops to – .39. There are, of course, other factors, particularly the age factor, which has not yet been considered. Still it is recalled that it takes a rather powerful factor to reduce a negative correlation of – .39 to zero or to make it positive. With reference to the age factor, it is interesting to note that the cities with the largest percentages of young women employed are also the cities with the smallest percentage of married women who are young, the correlation being – .43. This suggests that employment discourages early marriage. But we know that this negative correlation may be due to the sex ratio, for where there are excesses of women there are small percentages of women married and also large percentages employed. Holding the sex ratio factor constant, the correlation between the percentages of women employed and the percentages of young married women drops from – .43 to – .17, a low correlation, yet negative. There are, therefore, certain indications that employment of women may discourage marriage; but this interpretation of the negative coefficients is speculative. There may be other factors not yet considered, or other interpretations of these factors.

The conclusions are, then, that the great need is for better statistics of the family; that the method of standardizing the age distribution of the population for comparative purposes may be used to advantage with marriage statistics; and that the method of partial correlation can also be used profitably, provided small locality units be used and provided a careful consideration be given to a number of other factors. Even so, the interpretation of the meaning of the partial coefficients is not always convincing.

III

STATISTICAL STUDIES OF HEALTH AND MEDICAL CARE

HUGH CARTER

STATISTICAL methods have been applied so rapidly to problems of health and medical care during the past few years that any attempt to discuss all of them would be an extensive task. It seems more profitable here to examine the methods used in a few specific studies now under way. The Committee on the Cost of Medical Care[1] in its fact-finding activity is engaging in some twenty-five studies, and is co-operating with a number of other agencies which carry on research in the same field. All of these studies relate to the economics of health and medical care. A few of the problems which are being faced will be discussed in this paper.

The major factors in medical care are: (1) individuals, such as physicians; (2) institutions, such as hospitals; and (3) secondary therapeutic aids, such as drugs. Some of the methods of attack upon each of these will be considered. In all of the work in this general field the most elementary matters must receive careful and persistent attention. For example, in some fields of inquiry the definition of the statistical unit and the determination of the limits of variation to be allowed are simple matters, but in the problems under consideration they are surprisingly difficult.

[1] At the time this paper was prepared the writer was chief statistician of the Committee, while on leave from the Wharton School of Finance and Commerce of the University of Pennsylvania.

If the question is asked, "What is a doctor?" there is in the popular mind no difficulty in giving a definition. The same is true if one asks, "What is a specialist?" However, when one begins to enumerate the number of specialists in the United States, certain difficulties immediately arise. The directory of the American Medical Association, issued annually, divides physicians into three groups: general practitioners; wholly limited specialists; and physicians interested in, but not limited to, the practice of a specialty. This seems fairly straightaway. How does it come about that certain physicians are classified as specialists either "wholly" or "partially" limited? It is self-classification; the result of material furnished by the physician himself. When it is recalled that the specialist has greater prestige and receives higher income, it is clear that the material given in the directory does not exactly correspond with the facts. Sometimes a wish to become a specialist translates itself into a definite listing as such in the directory.

There are several ways in which this problem of defining the specialist can be approached. The first is training. One observes that the formal training of wholly limited specialists, beyond the basic general medical course, varies from prolonged and excellent work in post graduate courses and clinics, both here and abroad, to no formal training for the specialty. Another approach to the classification of specialists is the proportion of a physician's time devoted to work in his specialty. Many of the so-called wholly limited physicians devote only a portion of their time, in some cases less than half, to work in this field. Among the partially limited specialists the proportion of time given may be actually less than that given to some other professional activity. We know this from many statements by physicians regarding their own practice.

Still another approach to the problem of classification is through the various honorary societies to which a specialist is elected by his fellow-workers when, in their judgment, he has attained distinction in the given field. Such information is available in the medical directory and from the county societies. The difficulty, statistically, with this material is that the various groups of specialists have different standards for admission to membership. The problem is further complicated by the fact that the general practitioners are not a homogeneous group, as it is customary for such physicians to limit their practice. Many of them exclude surgery, work with venereal diseases, pediatrics, and other forms of professional work. It was apparent from an intensive study of a group of physicians in one community that there is a continuous gradation from the general practitioner, doing everything, to the wholly limited specialist who really is "wholly limited." Another complication grows out of the fact that only about two-thirds of the physicians are members of the American Medical Association, and information is quite fragmentary on the fifty thousand non-members.

As to the question, "What is a doctor?" it is necessary to define rigidly the limits of variation. For "doctors" range in ability and in training from close to zero to the highest rating one wishes to give; in age, from the early twenties to well into the eighties; in the amount of time devoted to their practice, from fifteen hours a day to practically no time at all. In any study of the relative merits of two forms of organization for rendering medical care, a basic question relates to the ability of the physicians in the two groups. One can observe off-hand a great difference in ability between the physicians in private practice in a given community and those in some Government service. This comparison of ability may favor one group or the other in a given instance, and may be due simply to historical accident; but in any event the relative ability of the two groups must be examined before any view can be taken as to the merits of the forms of organization involved.

The difficulties of defining the statistical unit are even greater in connection with most of the other factors in this field. What is a hospital? What is an illness? What is a drug store? The list could be lengthened indefinitely. In each instance the unit must be carefully defined, sometimes arbitrarily defined, before the analysis can proceed.

Since the Committee on the Cost of Medical Care is interested primarily in the cost item, a study of professional incomes and expenses is an important segment of the general problem. One observes that the public, faced with the rising costs of illness, has on occasion thrown the blame upon the physician and his supposedly ruinous fee schedules. The physicians, on the other hand, point to the rising cost of hospitalization and of innumerable expensive mechanical aids to diagnosis and treatment, such as the X-Ray and the cardiograph, made possible by the advance of science. Physicians call attention to the substantial amount of free work done, and maintain that their incomes are unreasonably low in view of the training and skill required. Since more heat than light has resulted from such controversies, it seems worth while to gather some facts as to professional incomes.

One of these income studies is localized in the Philadelphia area. An anonymous schedule was mailed to about twenty-five percent of the physicians in the community. It was felt that the natural and widespread reluctance of most individuals to reveal their incomes would be overcome in this way. Physicians who did not immediately reply to the schedules were followed up by using a familiar device. Each physician was requested, after he had sent in the schedules, to mail in a postal card indicating that he had done so. With the aid of a local committee of physicians and a follow-up program usable schedules were secured from nearly sixty percent of the physicians in the original sample.

This material is useful and is being carefully analyzed as to income and expenses, but the possibility of a biased sample requires attention. At one time the rapid flow of completed schedules coming into the Committee's office gave the field workers hope that detailed checking on the sample would not be necessary. At this stage of the work the percentage of returns established by the statistical office as desirable was set sufficiently high to reduce to small proportions the error growing out of failure to return the schedules. The original hope of large returns did not materialize and certain methods of checking were determined upon to give confidence in the final results. One of these checking devices is the familiar procedure used by the Bell Telephone Company in connection with its plans for future extensions. The technical staff of the telephone company finds a close relation, in a given section of the country, between home rentals and telephone usage. The rental value of homes and income is, of course, closely related. Through the courtesy of the technical staff of the telephone company, the Committee secured skilled assistance in estimating the rental value of each dwelling occupied by all the physicians on the lists. This included those replying and those not replying to requests for information. As this rental material comes to hand, it will provide valuable checking data. By superimposing the frequency distribution of rents in the groups of physicians replying and not replying, respectively, it will be possible to see something of the situation, even prior to the analysis of the two distributions. In this particular area, we can make rough estimates of income corresponding to the given rental. This will be useful, although, of course, rent is related to *total income* and we are primarily interested in *professional income.*

Other methods of checking the income data will include estimates by individuals and credit agencies well qualified to know the facts. The Committee wishes to feel sure that so far as this particular metropolitan area is concerned it will have a clear picture of doctors' incomes. What the situation is throughout the country, is being attacked in another study. It is, of course, more difficult for the physician than for many others to estimate his net income, particularly since most physicians are not given to detailed record-keeping. A physician seems to have a fairly good notion of his gross income, but his calculation of net income is as often as not erroneous. This statement can be made with confidence, since a certain itemization of expenses is contained on the schedule used in the studies. Few physicians have a clear notion of depreciation of equipment, sometimes a large item; as to the cost of supplies and the itemizing of various expenses, they are often quite hazy. However, by a detailed study of all available sources of information, we shall have reasonable confidence in the general outlines, at least, of the picture of professional incomes and expenses that is being built up.

An important institution in medical care is the hospital. People go to the hospital in increasing numbers, even for minor ailments; and as a consequence hospital costs have become increasingly important in the total cost of medical care. One may observe in passing that the hospital, through a curious evolution, is at present a strange medley—in part, a luxurious hotel for ailing individuals of large means; in part, a charitable institution for individuals unable to pay; and, in part, a laboratory available to medical schools for teaching purposes. The individual who is neither very rich nor very poor finds the modern hospital unsatisfactory. He is neither pauper nor millionaire. Sometimes he pretends to be a pauper, sometimes a millionaire, but in neither rôle is he entirely happy. In studying the hospital, the Committee on the Cost of Medical Care procured detailed information from several hundred institutions scattered over the country. It examined the general set-up of the organization and any attempts toward a rational adjustment of the institution to changing conditions. Field studies were made of the more interesting hospital innovations.

One obvious difficulty with the hospital is a lack of adequate cost accounting methods and general business control of the institution. The actual cost of particular services is unknown to the average hospital superintendent. On this matter of costs, the Committee is proceeding with an analysis under the direction of a cost accountant, after field work with a large number of institutions scattered over the country. This field work consisted of a personal inspection of the institution, gathering of all pertinent statistics bearing on costs within the institution, and discussion with the superintendent and others of the figures contained in annual reports and similar publications. One observes, in passing, that so far as the physical structure of the hospital is concerned the trend is toward a more rational type of structure to care for the sick of varying means. It is possible that studies under way will eventually contribute to better cost accounting.

A different problem is faced in analyzing the drug industry. One phase of this inquiry is a drug prescription analysis in a given community. Again, the interest is centered upon cost—the cost of prescriptions and the reasons for variations in cost. The analysis of the prescription slips filed with each local pharmacist presents certain practical difficulties, such as the general custom of recording the price in code. As a result, the tabulation and checking work goes forward slowly. For practical reasons, it was not feasible to establish a complete time series of changes in prescription costs over a period of years. We compromised by making a detailed analysis of costs in a considerable number of drug stores for the years 1929, 1919, and 1909.

The total number of ingredients entering into prescriptions is extremely large. There are the various drugs that the pharmacist speaks of as "official," since they are listed in the official pharmaceutical publications. These ingredients are required by law to conform to certain standards of purity and strength. There are also various drugs that are sold under a manufacturers' seal and a registered proprietary name. These "manufacturers' specialties" command a higher price than the official ingredients. The habits of the physician in the type of prescription written obviously affect cost. It is the custom of large pharmaceutical houses to send salesmen, "detail men," to visit the physicians. The detail man introduces new pharmaceuticals bearing the exclusive name of a manufacturer and attempts to convince the physician of the superior merits of these proprietary articles. Since there are a large number of such pharmaceutical houses, each with its detail men, it is clear that the problem of the individual pharmacist is greatly complicated by the necessity of stocking a large number of manufacturers' specialties, many of which move slowly.

In the analysis of prescriptions, attention was given to the cost of prescriptions of three types: those made up exclusively of manufacturers' specialties, those made up exclusively of "official" ingredients, and some combination of both of these types. It may be noted that the proportion of prescriptions containing manufacturers' specialties has risen sharply. It is the contention of many pharmacists that this is largely responsible for the rise in the cost of prescriptions. On the other hand, many physicians maintain that they have greater confidence in the materials compounded by large and estalished pharmaceutical manufacturers than in those compounded in the average drug store. These physicians state that such articles are compounded with greater care and are received in better condition by the patient since they are bottled and sealed by the original manufacturer. Certain physicians state that a visit behind the scenes in the usual drug store does not inspire confidence in the drug-compounding carried on there. We are not concerned at the moment with the merits of these varying points of view. Costs were considered by stores, prescription slips for analysis being selected at regular intervals throughout the twelve months' period. Another of the studies will consider the economics of the entire drug industry with a view to setting forth the salient facts.

One type of study being carried forward by the Committee is an attempt to see the entire field of health care and medical care in a given community. This is in a way the examination of a single case, though obviously such a "case" breaks down into familiar statistical problems. There have been an enormous number of segmental studies of health activities, such as the examination of the hospitals, the clinics, or the public health agencies in a given community. Such studies, valuable as they are, leave a great many questions in the air since one cannot understand a single group of agencies without understanding related agencies. In the Committee's attempts to see the situation as a whole, it is making intensive studies in selected communities. This procedure has been followed, for example, with the rural county of Shelby, Indiana.

A field worker and assistant go into the community, reside for several months, and compile all available data. The attempt is made to get a comprehensive picture of the people and of social and economic conditions. Detailed information on medical care is secured from every available source. This includes all of the private practitioners, both members of the regular medical profession and the so-called irregular practitioners such as the chiropractors, the advertising dentists and traveling medicine shows; it includes also detailed information from the public health agencies in the community, from drug stores, hospitals, etc. By such a procedure, the Committee attempts to see the situation as a whole in a way that would be impossible through unrelated segmental studies. Through coöperation with local medical agencies, through innumerable casual conversations with individuals in a position to know the situation, and through other means which occur to an alert and resourceful field worker, it is possible to get a detailed and illuminating picture of the medical care conditions. It is, of course, not presumed that such a picture is a typical one. Such a study is illuminating because of the confidence one has in the accuracy of the picture drawn. How far such conditions will be found to exist elsewhere, is a matter for speculation; only a large number of similar studies could determine that point.

Some of the sampling problems faced may be discussed briefly. One relates to a study of the total cost of illnesses for one year to some ten thousand families, scattered over the country. Since we are fundamentally interested in the cost of illness, this study is of importance. The attempt is made to get cost figures in detail. Obviously, ten thousand families are not sufficient in number to enable us to state with confidence what illness costs are in different parts of the country, nor how such costs are distributed within different economic groups in various sections of the country. This study will show how much ten thousand families spend for medical care, for what types of facilities they spend their money, and, in a general way, how these expenditures vary with economic conditions. Tabulations will be made for urban, rural, and small town areas; for those having hospitals and those lacking them.

One great handicap in studying problems relating to the economics of medicine is the lack of morbidity statistics. A reason for this lack is the common assumption that morbidity can be inferred, broadly, from mortality. That this is not true has long been evident. The work of the United States Public Health Service has gone far toward dispelling the fallacy. The wonder is that the notion still persists. The death rate from cancer, or from similar diseases which are extremely difficult to cure, may represent a large proportion of those suffering from the disease; but some other diseases, more widespread, rarely result in death. We shall never have the information desirable on the economics of health and disease until the fundamental morbidity statistics are gathered by official agencies. It is hoped that the study of morbidity in ten thousand families, along with other important studies in this field by other agencies, will hasten the arrival of comprehensive, official, morbidity statistics.

IV

STATISTICAL STUDIES OF DEPENDENCY

RALPH G. HURLIN

RELATIVELY little attention has been paid by statisticians to this long-recognized problem. Although the literature of dependency abounds with fragmentary statistics, they justify few conclusions. Here, as with many other social problems, the most urgent need relating to the use of statistics is still the primary one of obtaining data comprehensive and accurate enough to justify refined methods of analysis. It is well to emphasize at once that the collection itself of such statistics requires technical methods, and also that to obtain satisfactory dependency data presents large, though certainly not insurmountable, difficulties.

That there should now be so great lack of good quantitative data concerning dependency is somewhat surprising in view of the insistency of the problem and its large importance as a cause of governmental expense. In this country we are spending month by month millions of dollars in support of public dependents, and from its administrative aspect alone the problem deserves much more attention than it has received from the governmental statistical bureaus. Fortunately, there is now renewed interest in this subject, and we may hope that as a result the present deficiency of dependency data will not continue.

The Scope of Dependency Data

Narrowly defined, the problem of dependency is that of poor relief. Pauperism is a more descriptive, though less euphonious term, for we are not directly concerned with the normal dependence of non-earning on earning members within the family, although statistics on this aspect of the family correlated with those on family income will show much concerning the cause of poverty and, in turn, of dependency.

The statistics of dependency, strictly speaking, relate to dependence by persons on unearned support, obtained because of want, from sources outside the family. Such support may be partial or complete, temporary or continued. It may be relief in institutions; or it may be outdoor relief, that is, in the home of the person aided.

Less narrowly defined, the problem of dependency may be identified with that of poverty, thereby greatly enlarging the range of data with which we are concerned. The application of statistics to the larger problem will not be discussed here, except to point out that statistical data concerning poverty are also quite inadequate. We have a good deal of data on wealthiness, derived from the periodic income tax returns. But these returns throw little light where it is most needed, on the incomes of the large portion of the population with small resources. The studies which have been made of the cost of living show how families of moderate means spend their incomes; but they do not show what proportion of all families spend specified amounts. There is great need in this country for studies similar to those which have been made, and are now being made, in England under the direction of Professor Bowley, which show how much of the population exists below a specified subsistence level.[1] Such studies require exact statistical

[1] Bowley, A. L., and Hogg, Margaret H., *Has Poverty Diminished?* London, 1925, describes the results of the 1923–1924 studies and relates

technique. Since they cannot be made extensively, it is necessary that the sampling of families studied be entirely typical, in order that reliable comparisons may be made from place to place and from time to time. Great care is necessary also in defining the level below which poverty shall be recognized and in the recording of data on income and expenditures. Fortunately, in the English studies sound methods for such analyses have been demonstrated.

Within the range of dependency data, we may also include the entire body of social work statistics. This is appropriate, both because the present elaborate system of social work activities has grown out of the early organized efforts of poor relief, and also because in general the activities of social work organizations are still largely methods of assisting the poor; although in many instances beneficiaries may not receive any direct financial aid. The increasingly varied methods of modern social work present a very wide field for scientific and administrative use of statistical methods. Thus far this field is largely uncultivated. Space has been taken here to place social work statistics definitely within the scope of dependency data because social workers, not wishing to stigmatize their clients as dependents, tend to ignore the relationship.

Types of Problems Requiring Statistical Methods

The following paragraphs will illustrate briefly some types of problems in this field in the solution of which statistical methods are required.

1. *Measuring the volume of dependency.* For many

them to comparable findings for 1913–1914. In the New Survey of London Life and Labor, now under way, similar studies to ascertain the amount of poverty of a defined degree are being made under Professor Bowley's supervision.

purposes, it is desirable to know at a given time and for a given area how much dependency there is, either in terms of the number of recipients or cost of actual poor relief, or in terms of the number of beneficiaries of, or the cost of operating, all welfare or social work agencies. To obtain such a measurement involves, in the first place, consideration of the original record-keeping of the organizations concerned. This is primarily a matter of accounting, and nowhere are standardized accounting methods more needed than in these non-profit making, publicly supported organizations.

The combination of original records of various organizations is distinctly a statistical problem. It is one which requires much ingenuity. Assuming that there has been reasonable attention to record-keeping, it will not be difficult to obtain an accurate count of inmates in institutions at any time, and their classification with respect to age, race, and other social attributes, should also be easily possible. There is, however, great lack of uniformity and of adequate detail in our statistics of institutional population. It is also a straightforward matter to count the recipients of outdoor poor relief. In the latter case, however, difficulty arises because aid is given to cover varying lengths of time so that it is not entirely easy to obtain a count of persons receiving aid at a given time. Since the usual period for which aid is given is not longer than one month, a month appears to be a good interval over which to count outdoor relief recipients. It is much less easy to count all the beneficiaries of social work organizations, because the services they receive vary *continuously* from outstanding assistance at one extreme to very brief attention, or only cursory advice, at the other. Under these circumstances, appropriate terms and definitions must be carefully devised, adopted, and actually put into use, before great reliance can be placed on such data for the purpose of comparison or combination.

Determining the cost of all organized efforts to alleviate distress should be somewhat easier than determining the number of beneficiaries of these efforts. Here, also, uniform accounting methods are needed. In aggregating costs, we are not confronted by the problem of duplication of aid to the same persons by different agencies engaged in the same or different types of work. In general, it is much easier to measure the extent of service in terms of persons served for individual types of welfare activity than for all types combined, but it would be of interest, none the less, to know what part of the total population is in receipt of assistance from any type of organized welfare agency. Noteworthy progress in this direction is now being made in some thirty-two cities by the Committee on Registration of Social Statistics,[2] at the instance of the National Association of Community Chests and Councils. Thus far its results have been prepared and published only for individual fields of activity, but its objective is to compile complete statistics of welfare activities for the localities studied.

Estimates of the total bill for dependency are frequently made, but all of which the writer knows must be regarded as only crude approximations. Probably the most careful attempt to approximate total cost of benevolent enterprises in this country, of recent date at least, is that of Paul H. and Dorothy Douglas in

[2] This is a joint committee, jointly financed by the Committee on Local Community Research of the University of Chicago and the Association of Community Chests and Councils.

their essay "What Can A Man Afford?"[3] Their estimate was made for the year 1921 and was presented as a "philanthropic budget," or statement of amounts needed for reasonably adequate support of existent objects of philanthropy, rather than an estimate of what was actually spent in that year. The estimate for institutional care of dependents, outdoor relief and service, hospital and dispensary care, and public health nursing, for that year, was $342,000,000. The figures of the Committee on Registration of Social Statistics for only thirty-two cities will give a much better basis for estimating some of these items than was available when Professor and Mrs. Douglas made their estimate.

During the past year the Russell Sage Foundation has been collecting regular monthly statistics of outdoor relief by public and private relief-giving agencies in cities of over 115,000 population in the United States and Canada. We have used a very simple schedule, asking only for the number of families receiving financial aid in their own homes and the amount of aid given during each month. There has been good response to our request for these data and well over three hundred such agencies in seventy-eight cities (seventy-five in the United States and three in Canada) are now contributing these monthly data. The figures for each month are compiled during the succeeding month, and are thus available for current indication of changes in the volume of outdoor relief in the larger cities. Because these data are new, the accompanying table is presented. It may be noted that the 159,601 families reported as receiving outdoor relief during the month (although there is probably some small amount of duplication) probably

[3] *American Economic Review*, XI, No. 4, December, 1921.

ANALYSIS OF RELIEF OPERATIONS IN DECEMBER, 1929, OF 337 AGENCIES IN 78 CITIES OF OVER 115,000 POPULATION

Summary by Type of Agency

	All Agencies	Public Departments			Family Case Work Societies			Salvation Army	Other Agencies
		Poor Relief	Mothers' Aid	Veterans' Relief	Non-Sectarian	Jewish	Catholic		
All agencies reporting:									
Agencies reporting	337†	42	44	15	64	47	25	56	51
Families aided	159,601	54,689	29,107	6,087	28,918	5,387	4,390	25,090*	5,933
Relief given	$3,878,082	$1,415,690	$1,292,730	$195,167	$496,941	$207,550	$78,143	$106,459*	$85,402
Percent of total:									
Families aided	100	34.3	18.2	3.8	18.1	3.4	2.8	15.7*	3.7
Relief given	100	36.5	33.3	5.0	12.8	5.4	2.0	2.8*	2.2
Agencies in 36 cities from which supposedly complete reports were received:									
Agencies reporting	193†	29	26	9	31	28	15	29	29
Families aided	99,405	38,467	20,763	4,282	13,172	2,947	2,934	14,036*	2,804
Relief given	$2,704,954	$1,111,238	$932,000	$151,908	$252,570	$108,240	$65,631	$56,305*	$27,062
Percent of total:									
Families aided	100	38.7	20.9	4.3	13.2	3.0	3.0	14.1*	2.8
Relief given	100	41.1	34.5	5.6	9.3	4.0	2.4	2.1*	1.0

† Smaller than sum of agencies in following columns, because figures of some public departments are classified by type of relief under more than one heading.

* Relief work of Salvation Army abnormally high in December on account of extensive Christmas giving.

represent as many as 650,000 persons, or nearly the equivalent of the population of so large a city as Pittsburgh.

2. *Computing Rates of Dependency.* In making either geographical or temporal comparisons of the extent of dependency, it is necessary to consider, in addition to the annoying question of adequacy of the actual figures, other factors affecting comparison, especially differences in the number of persons at risk of dependency. Rates to population are therefore needed. It is not necessary to dwell on the familiar liabilities of ratios, but they need to be taken into consideration.

It is probable that dependency rates vary both with race and age composition of the population. Immigration has been supposed to raise dependency rates, and Negroes being economically handicapped probably contribute more than their proportional share of dependents, notwithstanding a generally lower standard of relief-giving to Negroes than to whites. Both age and race factors are undoubtedly of different importance for different classes of dependency.

We have so little adequate information concerning crude dependency rates that it is not surprising that there is almost nothing in the way of specific rates for racial and age groups. The much-criticized study of the biological aspects of immigration made for the Committee on Immigration and Naturalization of the House of Representatives of the Sixty-seventh Congress attempted to determine specific rates for various types of social inadequacy, including dependency, for each important nationality group in our population, as a guide to the Committee in determining immigration policies.[4] This study illustrates an important

[4] *Analysis of America's Modern Melting Pot—The Determination of*

occasion for the use of statistical procedure. The results, however, were probably misleading, because of limitations of the data on pauperism, crime, and other types of inadequacy, obtained for the various racial groups, and also because in determining the rates no allowance was made for either age or sex differences in the populations at risk.

3. *Measuring Trends and Fluctuations in Dependency.* Determining consecutive changes in the volume of dependency over short intervals for study of fluctuation and trends is useful, both for what it shows concerning dependency and for what is implied concerning changes in the status of the large section of the population below or near the poverty line. Lacking comprehensive data, recourse must be had for such studies to samples presumed to be representative. A number of attempts have been made to produce significant dependency indexes. I. M. Rubinow prepared a dependency index for the city of New York covering the four-year period 1914–1917,[5] and F. Stuart Chapin has experimented with a similar index for Minneapolis.[6] Maurice B. Hexter has also worked with an index of dependency for Boston, and similar attempts have been made in other places.

The procedure for construction of many economic indexes has been to combine several related but distinct series of data, each supposed to reflect fluctuations in the phenomenon under consideration. This was the method of Rubinow's early index.

the Rate of Occurrence of the Several Definite Types of Social Inadequacy in Each of the Several Present Immigrant and Native Population Groups in the United States. Washington: Government Printing Office, 1923.

[5] "A Dependency Index for New York City," *American Economic Review*, VIII, 1918, pp. 713–40.

[6] "A Dependency Index for Minneapolis," *Publications of the American Sociological Society*, XIX, 1924, pp. 200–2. Also, *Social Forces*, V, 1926, pp. 215–24.

To construct such an index with precision is, of course, a time-consuming task, for attention must be given to adjustments of the individual series for differences in amplitude of fluctuation, timing of fluctuation, underlying trend, and relative importance or weight. Like the selection of the series to be combined, all these adjustments depend on the judgment of the investigator, and introduce important possibilities for distortion of results. In general, it seems to the writer that an index of dependency is likely to be more useful, if only because more easily interpreted, the more homogeneous the data from which it is constructed. It is evident that in the case of dependency series a great deal of ground work is needed in developing sound elementary series of data and discovering their properties, before very valuable composite indexes can be produced.[7]

During the past four years we have been making an intensive study of the data of family welfare agencies. With their coöperation, we now have the beginning of a group of homogeneous series of monthly data, which in time, we hope, may be valuable for index purposes. The accompanying diagram illustrates the relationship of curves resulting from the combination of three sets of these data: (*a*) the number of active cases; (*b*) the number of cases taken up; and (*c*) the number of cases granted relief. Thirty-seven large city agencies are represented over the four-year period. These are crude indexes only, obtained by selecting for each month the median of relatives computed to the base period shown. Adjustment has not been made for differences in amplitude of fluctuation, and seasonal fluctuation has not been eliminated. Nor has adjustment been made for differences in timing of the

[7] Cf. the discussion of social indexes in Chapter I.

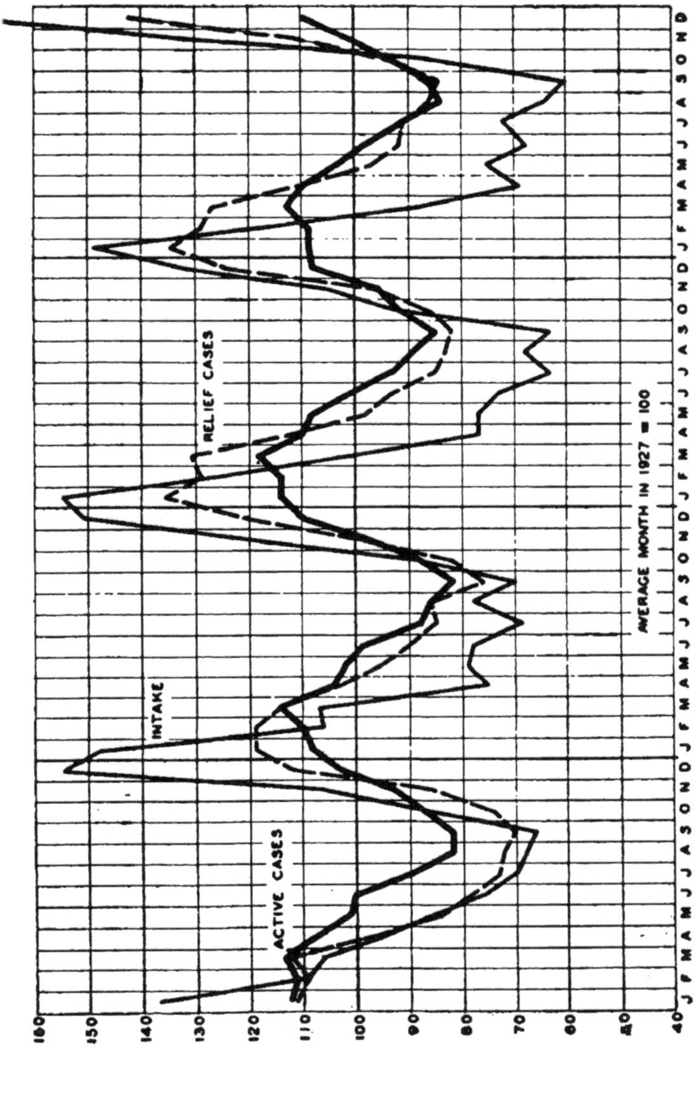

THREE INDEXES OF CHANGE IN CASE LOADS OF FAMILY WELFARE AGENCIES

seasonal fluctuation. Failure to make the latter adjustment reduces materially the amount of seasonal fluctuation shown by the index, thus indicating for the group of agencies considerably less than the average variation between summer and winter. It will be noted that the three curves show striking regularity in seasonal fluctuation, which, it is assumed, is some indication of goodness of the data. The higher points in the intake and the relief curves in December, 1929, coincide, we think, with increase in unemployment.

4. *Norms for Comparison of Administrative Practices.* From the systematic collection and analysis of comparable data for welfare agencies, such as is being made by the Committee on Registration of Social Statistics and by the Russell Sage Foundation on a more limited scale, are derived average measures of numerous aspects of social work practice. These measures are available as standards for comparison with similar measures pertaining to individual agencies. Such averages are not, of course, in any sense standards of efficiency or excellence. They are rather points from which to measure variability, but as such they have considerable practical value in pointing out extreme variations in practice which deserve examination.

During the four years in which we have collected comparative figures from the family case work field, we have computed many of these averages month after month. They have been of value both in describing the practice of the group of agencies as a whole and for the diagnostic purpose just indicated. Some of these averages, determined from the data of forty-two agencies for the year 1928, are as follows:

Percent of active major care cases each month given relief	44
Percent of active major care cases each month newly taken up	14
Ratio of client interviews outside office to each one in office	1.5
Average amount of relief per month to cases receiving relief regularly	$32.34
Active major care cases each month per worker dealing with cases	30

5. *Scales for Measuring Status of Clients or Quality of Work.* An important problem which is beginning to receive attention in this field is the invention of devices for measuring complex attributes of clients, workers, or agencies. Case-working agencies are familiar with the use of psychological tests in diagnosing the mental condition of clients. Another rating scale of this sort is F. S. Chapin's dwelling-house score card.[8] The budget sheets used by many case-working agencies for determining family income and need of relief suggests an opportunity for development of a standardized measuring scale, which would prove of value not only for the immediate purpose of diagnosing need of relief, but also for correlating amount and frequency of relief given with disabilities of the family.

An interesting attempt is now being made under the direction of Dr. Ellery F. Reed, of the Trounstine Foundation, to develop an instrument for measuring the efficiency of work of family case work agencies. A schedule has been devised for recording the status of the case at the time of application, the work done by the agency, and the results observed, with respect to each problem presented by the case. These records

[8] "The Quantitative Scale for Rating the Home and Social Environment of Middle Class Families in an Urban Community," *Journal of Educational Psychology*, XIX, 1928, pp. 99–111.

are made in quantitative terms, and when combined yield coefficients which constitute the ratings or scores for the case. The plan calls for the application of the test to a sample of the cases of each agency compared, and the average of the scores obtained gives the rating for the agency. In view of the variability of situations treated by these agencies, there is obviously an important problem of sampling here. There are also troublesome factors presented by inequality in the recognition of problems in different cases and in the assignment of responsibility for the outcome of situations treated. Painstaking work has been done in testing and revising this scale and it is now being used for comparing the Cincinnati case-working agencies. There is not yet evidence that a high degree of reliability can be expected from this scale, though this may be demonstrated. If its results should prove trustworthy, the great labor in applying it will not prevent its wide use, for some objective test for results in social work is in great demand. Whether or not the Reed method proves satisfactory as a tool of statistical measurement, it has already demonstrated its utility as an aid in finding weaknesses in the practice of agencies, and it undoubtedly represents a step toward measurement in this field.

6. *Analysis of Causes of Dependency.* A decade or two ago the classification of dependents by alleged causes of dependency was much in vogue. The futility of such classification has become so generally recognized that now many social agencies make no attempts to assign the causes of dependency among their clients. Instead, they wisely place emphasis on the factors in the situations dealt with which are relevant to treatment. Ascertaining the causes of dependency is, however, a task which needs attention, and because

causes change repeated study is needed. Correlation techniques, rather than crude classifications of persons aided with respect to apparent reasons for want, are, of course, required. It is perhaps worth emphasizing here that the immediate cause of dependency is the willingness of some person or agency to assist those in need. The causes of which knowledge is most needed are those of the poverty or destitution lying back of dependency. This raises the question as to whether or not, in studying causes of need as exhibited by dependents, enough attention has been given to the fact that a selected group, not necessarily typical of the larger group in which we should be more interested, is being dealt with.

For dependents, however, there is possibility of ascertaining detailed accurate data as to causative factors much more readily and frequently than is the case with the poor who are not dependent. It is reasonable, then, to prosecute the search for causes among dependents. In this connection, it should be pointed out that many types of control data for the general population are needed. One method of investigating causes of dependency is that of correlating fluctuations in dependency series with those in correlative social series. Chapin found for Minneapolis no better data for testing the relation of dependency to general economic conditions than bank clearings, and discovered that the bank clearings index for Minneapolis gave a lower correlation with Minneapolis dependency than a similar index for the country as a whole. For his purposes, an index registering changes in the economic status of the wage-earning population in Minneapolis, such as a comprehensive index of employment, unemployment, or volume of wage payments, would have been in order, but this was lacking.

The Outlook for Improved Dependency Statistics

The most promising sign for improvement of dependency data in this country at present is the item in the Appropriation Bill now before Congress, which will, if approved, permit the Federal Children's Bureau to assume the work which the Committee on Registration of Social Statistics is now carrying on. The Children's Bureau already has to its credit the inauguration of standardized statistics for juvenile courts throughout the country, and is developing countrywide statistics of the issuance of work permits for children of school age. While not perhaps the most logical bureau in which to place the collection of general dependency statistics, it is certain that these auspices will insure intelligent interest in these statistics. It is to be hoped that with Government funds available this collection of data may be rapidly expanded. As has been pointed out already, these data are collected for individual cities on a community-wide basis. While this basis of collection is logical and important it is greatly to be desired, since much of welfare administration is under state control, that state governments be encouraged and assisted to develop comprehensive and uniform state dependency statistics.

In conclusion, the writer would like to emphasize the point that good dependency data cannot be produced by command. Something more is required than merely to devise good forms and send them forth to be filled in. It is possible, however, to encourage welfare agencies to keep and to use good statistics. These data will improve in direct proportion to the extent to which they are used by the agencies producing them. It is, therefore, important in devising any plan for good statistics of welfare work to have in mind the possible usefulness of the data required to the agencies which must compile them.

V

STATISTICAL STUDIES OF RACE RELATIONS

DONALD YOUNG

THE statistician who applies himself to race relations is usually seeking an answer to some aspect of one or the other of two questions. First, what is the relative status of the groups under consideration? Second, why do differences in status between racial groups exist? Both questions have immediate value in the development of social programs; both contribute to the multitude of scientific hypotheses and theories out of which may grow some ultimate understanding of the processes involved.[1]

If the investigator is interested in questions of the first type, his task will not present any novel problems. His work will be concerned chiefly with gathering data and working them up into significant form. Given the problem of determining the extent to which crime is found among European immigrants having a particular racial inheritance, the difficulties in establishing the criminal status of this group would differ little in principle from those involved in an analysis of criminality throughout the entire population. The physical task of collecting data and the

[1] Too often the statistical method is regarded as the only tool which gives real promise of effective use in building up a knowledge of race relations. For the possibilities of a quite different approach see Thomas, W. I., and Znaniecki, F., *The Polish Peasant in Europe and America*, 2nd edition. New York: Alfred A. Knopf, 1927. Vol. I, pp. xv + 1115; Vol. II, pp. vi + 1135.

statistical effort necessary to make them usable might be enormous, but the fact that race relations were involved would be of no methodological importance, except in limiting the field of study. Since the major obstacles in the study of racial status are not peculiar to research in race relations, they need no detailed consideration here.

On the other hand, attempts to find out why differences in racial status exist are met with difficulties inherent in the nature of the subject matter. Let us assume that we know the extent of criminality among the European immigrants that we set out to study. Let us assume, further, that there is more criminality among them than among the native citizens of old American stock. How can we then determine the origin of this differential? Here it is not enough to be familiar with the sources of material, the limitations of questionnaire and schedule, and the proper use of formulæ and charts. The concept of race itself now becomes a factor of such importance as to invalidate the results of research projects which have emphasized mechanical statistical method and minimized consideration of the assumptions underlying all comparisons of racial groups.

It is a truism to say that what a man accomplishes is attributable to his heredity and his environment. When a European immigrant becomes a criminal, these two sets of factors must have made him so. But how can we find out the relative influence of his blood and culture? The task of discovering that he *is* a criminal is indeed child's play compared with that of answering *why* in specific terms of race *or* culture. The accuracy with which any such answer can be made depends, in the first instance, on the validity of the assumptions which all racial statisticians must make before they may quantitatively attack their problems. It is because these assumptions are inseparably linked with the unique phenomena of race, while the other statistical difficulties are not, that the former alone will here receive attention.

The statistical approach to the study of race relations is handicapped at the start by the difficulty encountered in any attempt to divide humanity into authentic biological sub-groups. To constitute a race, people must have common innate characteristics because of common ancestry. Individuals of simple racial origin with respect to recent generations may be found almost anywhere; but whole communities free from intermixture are practically non-existent. Consequently, the basic statistical units compared must, as a rule, be so loosely defined as to be meaningless. The alternative is the adoption of a cultural standard for discrimination, quite different in its implications from one of race.

This holds true for the entire world, if a few isolated districts are left out of consideration. The omission of such districts is perfectly proper, since race relations are of little concern where but one race exists. In the United States, white men from all parts of Europe have mixed with red and black to such an extent that full-blooded Indians east of the Mississippi, and Negroes of pure African stock anywhere in the United States, are now in the minority of their respective, so-called, races.[2] Orientals, themselves of many origins and dissimilar characteristics, show ineffective re-

[2] Herskovitz, M. J., *The American Negro, A Study in Racial Crossing*. New York: Alfred A. Knopf, 1928. Pp. xiv + 92. There is no better small volume than this from which to gain a knowledge of Negro mixture with white and Indian in the United States and an understanding of the statistical problems involved in its study.

luctance to continue the process of intermixture when in contact with other peoples.[3] Negroes in Africa do not all have flat noses, dark skins and kinky hair. The countless invasions and migrations which have swept over all parts of Europe from the east and south have left their impress on the bodies of the present generation.[4] What chance has the statistician of unraveling such a tangle? He must unravel it, at least in part, if he is to study race relations with due consideration for the biological elements involved.

It is a practical impossibility to find groups of individuals who can be definitely classified as racially pure members of a specific race, biologically defined, under conditions where the study of race relations is feasible. Since this is the case, interpretations of data in terms of race alone should be accepted with extreme caution. Dr. H. H. Laughlin's noteworthy analysis of immigrant defectives, for example, is a study of transplanted nationalities and cultures, and consequently of little worth in racial comparisons; but it is frequently so used.[5] Studies of colored

[3] For discussions of Oriental race relations in the United States, see Coolidge, M. R., *Chinese Immigration*. New York: Henry Holt & Co., 1909. Pp. x + 531; Millis, H. A., *The Japanese Problem in the United States*. New York: The Macmillan Co., 1915. Pp. xxi + 334; Mears, E. G., *Resident Orientals on the American Pacific Coast*. Chicago: University of Chicago Press, 1928. Pp. xvi + 545.

[4] See Ripley, W. J., *The Races of Europe*, a sociological study accompanied by a bibliography of the anthropology and ethnology of Europe, published by the Library of the City of Boston. New York: D. Appleton & Co., 1899. Pp. xxxii + 624.

[5] Laughlin, H. H., *Analysis of America's Melting Pot*. Statement before the House of Representatives, Committee on Immigration and Naturalization, Serial 7C, November 21, 1922; Laughlin, H. H., Statements before the House of Representatives, Committee on Immigration and Naturalization, Serials 5A, March 8, 1924, and 1B, December 10–16, 1924.

The criticism of Dr. Laughlin's original study which he himself

peoples are even more likely to be accepted as real racial data, regardless of race crossing and ever present culture conflict. In view of the almost universal racial intermixture and differential cultural development found in modern heterogeneous civilizations, it is suggested that the student of race relations should consciously direct attention away from the biological concept of race and deliberately focus it upon cultural groupings.

Biological scientists should, of course, continue their efforts to uncover significant differences between groups.[6] No criticism of their praiseworthy efforts to understand and classify the peoples of the world is here intended, unless it be of their emphasis on differences rather than upon the more important similarities, long popularly neglected. The fault with which we are here concerned lies in the misuse of their findings by overanxious social scientists, too eager to explain achievement and group conflict in terms of racial qualities.

The ultimate significance of racial characteristics depends on their relation to cultural development and on their permanence. Anthropologists make no claim that pigmentation, hair texture, bodily proportions, and similar racial features, can now be shown to have a causal relationship with culture. Head form and characteristics affecting the brain and nervous system would seem to be more closely related to success in the modern struggle for existence, although there is

has considered most seriously is that by Jennings, H. S., "Undesirable Aliens," *The Survey*, December 15, 1923.

[6] For a bibliography of the contributions of the physical anthropologists see Haddon, A. C., *The Races of Man and Their Distribution*. New York: The Macmillan Co., 1925. Pp. 173–85; Martin, R., *Lehrbuch der Anthropologie in systematischer darstellung*, 2nd edition. Jena: Gustav Fischer, 1928. Vol. III, pp. 1184–1816.

more skepticism about this now than existed thirty years ago. We are not even sure, for illustration, that the cephalic index necessarily remains unchanged through generations. Attempts to correlate head shape, brain size, and weight with achievement have failed to give us uniform results. The cultural factor in disease susceptibility and immunity is increasingly apparent. What right, then, has the theorist to utilize such material in support of racial doctrines?

Race relations, after all, are cultural phenomena, primarily culturally determined. Race conflict, for example, is more closely correlated with economic competition than it is with blood.[7] In the United States friendliness to the Oriental immigrants has varied in relation to their numbers and their occupations. The Indian has been an outcast beneath contempt; but he has also been a source of pride to the socially elect who have boasted of Indian blood flowing in their veins.[8] The Nordic Swedes have not always been respected in the United States, nor have

[7] For an illustration of an extensive and valuable study of race relations between Europeans and American Indians, Pacific Islanders and Africans, using the method of comparative anthropology, with the point of attack entirely cultural and the emphasis on economic competition, see, Muntz, E. E., *Race Contact.* New York: The Century Co., 1927. Pp. xiv + 407.

[8] The following volumes are among the best sources for comprehensive information concerning race relations between the American Indian and the European: MacLeod, W. C., *The American Indian Frontier.* New York: Alfred A. Knopf, 1928. Pp. xxiii + 598; Meriam, Lewis, and Associates, *The Problem of Indian Administration.* Institute for Government Research, Studies in Administration. Baltimore: The Johns Hopkins Press, 1928. Pp. xxii + 872; Schmeckebier, L. F., *The Office of Indian Affairs, Its History, Activities and Organization.* Institute for Government Research, Service Monographs of the United States Government, No. 48. Baltimore: The Johns Hopkins Press, 1927; Lindquist, G. E. E., *The Red Man in the United States, An Intimate Study of the Social, Economic and Religious Life of the American Indian.* New York: George H. Doran Co., 1923. Pp. xxviii + 461.

the Germans, Irish, and other Europeans of varying white ancestry.[9] In Europe, too, this thing we misname race relations has not been constant in its peculiar manifestations.

Hence, to initiate statistical analysis with supposed biological groups as units for comparison can lead only to confusion, unless biological and cultural groups chance to coincide. This is true only occasionally, although undoubtedly there is a great degree of overlapping. This overlapping, frequent but rarely complete, affords partial validity to the results of numerous studies founded on the false assumption of coincidence between actual and alleged racial origin. It is doubtful, however, whether the accidental accuracy of such studies affords scientific justification for the acceptance of a basic false hypothesis.

So long as the term "race relations" is used to designate social phenomena which are only in part correlated with ancestry, confusion must remain. "Cultural relations" is also a misleading term, for it neglects the possible biological factors. Some new phrase should be devised which will keep calling attention to the fact that the problems of race relations are of many kinds and spring from many causes. Ancestry may be important, but so are other things. Meanwhile, we shall, for want of a more accurate phrase, continue to speak of race relations, trusting that the broader meaning here given to the concept will not be overlooked.

As his unit of study, the statistician may use almost

[9] See, Abbott, Edith, *Immigration: Select Documents and Case Records*. Chicago: University of Chicago Press, 1924. Pp. xxii + 809. Also *Historical Aspects of the Immigration Problem: Select Documents*. Chicago: University of Chicago Press, 1926. Pp. xx + 881. Stephenson, G. M., *History of American Immigration, 1920–1924*. Boston: Ginn & Co., 1926. Pp. vi + 316.

any group of people who have "high visibility."[10] The latter may be due to color, apparent slant of eyes, hair cross section, language handicap, dress peculiarities, shape of nose, or any other differentiating feature, superficial or profound. It may be cultural or ancestral, or a combination of the two. It may be everlasting or ephemeral. The biological scientists have given us no firmer basis for discrimination, and with increasing race mixture they are not likely to do so for some time to come. What is more, the man on the street does not care. So long as he is able to make distinctions and miscall them racial, the social phenomena which we call race relations will continue, and should be studied as they are, and not as they ought to be in a scientifically organized society.

Roughly classified, the statistical approach has been of value in the study of race relations through its application in four overlapping fields: (1) The technique of measuring and comparing the physical characteristics, primarily external, of races has been developed into an art requiring no mean skill. (2) Cultural achievement of races is also subject to statistical methods, although this approach has unfortunately been neglected, due possibly to emphasis on the customary methodology of the historians. (3) Recent developments in the measurement of psychological characteristics have opened up a fertile field for further exploration. (4) In this last field might be included the analysis of social attitudes and opinions, but since this subject is discussed in a later chapter by Professor Rice it seems advisable to place it in a special category here and to omit further discussion of its possibilities.

[10] This term is borrowed indirectly from Professor R. E. Park's lectures at the University of Chicago.

1. External physical characteristics were the first factors in race relations to receive statistical treatment in the modern style. Such differences were most striking and most easily classified and measured. They seemed to be permanent because of racial heredity, and they also appeared to be significant in that there was an apparent rough correlation between many of them and accomplishment. The extent of their importance, if not their permanence, may now be questioned.[11]

It is, of course, interesting to the social statistician —and in some ways valuable for him—to know how peoples of the world differ in appearance and physical structure. Light may be thrown upon many questions by examination of the varying physical types of man. In view, however, of the extent of present-day artificial living conditions, the interpretation of such variations is becoming increasingly difficult. To the social scientist, furthermore, physical characteristics of races are proper data only if, first, they can be consistently correlated with the social activities and potentialities of the group possessing them; or, second, if they act as stimuli influencing the activities of some other group.

So far as outward physical characteristics are concerned, we may say that although their correlation with inherent limitations on cultural activities has

[11] See Kelsey, Carl, *The Physical Basis of Society*, 2nd edition. New York: D. Appleton & Co., 1928. Pp. xxii + 526. In Chapter XIV, Dr. Kelsey has concisely summarized and evaluated our present knowledge of racial differences in so far as they may throw light upon the development of society. The statistician should become thoroughly familiar with this type of material before attacking a problem involving a causal explanation of the relation between race and any phase of cultural development. Too often the research expert has been so intent on statistical technique that he has forgotten the necessity for understanding the subject matter to which his formulae are applied.

long been popularly assumed, no such causal connection has ever been scientifically demonstrated to any reasonably significant degree. The common belief that achievement is an accurate indication of potentiality is in a large measure responsible for the easy acceptance of this assumption. For example, yellow skinned people are thought not to have achieved the same complexity of material culture as those with a white skin, blue eyes. and blond hair. It is, of course, possible that such a correlation may be demonstrated in the future, but a new technique seems to be required before this assumption can be accepted as a basis for further investigation. It is perhaps regrettable that the problem is not so simple as in the case of cattle, where easy experiment and simple statistical method have established a direct relationship between hair color and milk production potentialities. A relationship between hair or skin color and possibilities in the way of material or spiritual improvement in human beings is still in the realm of idle speculation.

However, there is no doubt that external physical racial characteristics act as stimuli which directly influence the social activities of dissimilar groups.[12] The ordinary southern white man in the United States unquestionably reacts in many ways to the color of a Negro's skin. The outward appearance of an Oriental, regardless of his other qualities, may be the immediate cause of his exclusion from a community in any one of the Pacific Coast states. It may even lead to a lynching without possible justification in anyone's conduct. Red hair may be a handicap or advantage

[12] For illustrations of this fact, see Lasker, Bruno, *Race Attitudes in Children.* New York: Henry Holt and Co., 1929. Pp. xvi + 394; Bogardus, E. S., *Immigration and Race Attitudes.* New York: D. C. Heath & Co., 1928. Pp. xi + 268.

in the social contacts of its possessor. In this respect, at least, racial features must not be neglected by the social statistician.

At this point a further common assumption should be questioned. It must be remembered that no universal response to any given racial characteristic has been established. White communities have paid bounties on the scalps of Indian women innocent of evil. At other times or places they have taken pride in intermarriage with members of that race. Parts of continental Europe have long been known as places where race feeling against the Negro was relatively weak. But now that Negroes have become more frequent visitors, and white Americans and South Africans have told their story, some of these same Europeans react quite differently than before the World War. Jewish features may cause one response in the Gentile New Yorker and another in an isolated Pennsylvania German villager. Since this is true, no generalizations which would hold if time or place or individuals were changed can be made from studies in this limited field.

On the whole, it would seem that the student of race relations who devotes much of his time to statistical analysis of external physical characteristics, as such, is wasting effort in a field which should at present be reserved for the biologist, the psychologist and the physical anthropologist. Their work is as yet far from completion. True, they have established differences; but a knowledge of the importance of these differences is still wanting. In any event, a technique for the use of external physical characteristics as social data, except possibly as stimuli which call forth reactions from other groups, has yet to be devised.

2. Students of race relations have made use of cultural achievement in two major ways.[13] The first of these, historically, has been the comparison of two or more racial groups, usually geographically separated, as superior and inferior civilizations; with the inference or direct statement that variations in cultural levels were due largely to inborn limitations and advantages.

There is almost no limit to the number and extent of studies of this type. Orient and Occident have been contrasted, as have innumerable small portions of each, with the end in view of establishing the nature of variations, together with an interpretation of their origin. As a rule, during the last century or two, it has been taken for granted that some, if not all, of the variations could, *a priori*, be attributed to innate differences between Orientals and Occidentals. Of course, there have also been interpretations which were almost entirely economic, geographical, or historical; but it was a bold author who left out all reference to inherent dissimilarities between Asiatics and Europeans. It is safe to say that no people on earth has been left untouched by this type of study. It is also probably safe to say that the results in terms of racial qualities have been purely negative.

[13] Three books which comprehensively and effectively discuss the relation between race and cultural achievement are: Pittard, Eugene, *Race and History, An Ethnological Introduction to History.* New York: Alfred A. Knopf, 1926. Pp. xxiii + 505; Hertz, Friedrich, *Race and Civilization.* New York: The Macmillan Co., 1928. Pp. ix + 328; Hankins, Frank H., *The Racial Basis of Civilization, A Critique of the Nordic Doctrine.* New York: Alfred A. Knopf, 1926. Pp. ix + 384. There are, of course, literally hundreds of other works published in the last century dealing with the same question, a large proportion of which contain some statistical material. The most important of them can be traced through the three references above. A major reason for selecting Pittard, Hertz, and Hankins for mention here is their critical attitude toward the common assumptions underlying racial comparisons.

As a rule, these comparisons have been of material, intellectual, or sometimes even "spiritual" achievements, all of which resist measurement in terms of higher and lower standards. Qualitative rather than quantitative judgments have been liberally sprinkled over the lot. It is a rare historian who utilizes statistics when he analyzes civilizations other than his own. The cultural anthropologists stand almost alone in the avoidance of qualitative comparisons, although an occasional social historian has departed from the age-old traditions of his fellow craftsmen in refusing to interpret another culture in subjective terms. An accepted concept of progress, as yet not available, will have to be provided to validate this method of evaluating peoples through their cultures. Even should racial characteristics and culture show an almost perfect correlation in an unlimited number of cases, one would hardly be justified in assuming a causal relationship between the two, unless all other factors were held constant. This in itself constitutes a condition impossible to satisfy.

Geographic environment, cultural setting, and the passage of time, are three elements in the situation which at present appear impossible of reduction to a common basis as between peoples separated by space, history, or date. Who can say what the pre-Columbian American Indian might have achieved had he possessed horses and cattle? What if he had been able to borrow the concept of the wheel and its many uses from some other race? What if he had been left alone for only a few additional centuries, unhampered by colonists, conquistadores, Christian missionaries, rum and rifles? What if he had built upon Hebrew theology, Greek thought, and Roman law?[14] Isolation as well as other physical handicaps must be taken into account. Since we do not know how to include these things in our statistical reckoning, must we therefore assume that their weight is trifling compared with that of blood?

More useful has been the analysis by the statistical method of modern social problems, cultural in nature but with a racial differential. Crime, housing, illiteracy and educational retardation, family maladjustment, poverty, disease, and many other phases of social inadequacy, show variations in degree, if not in kind, when racial groups are compared. American Negroes have more illegitimacy than the native whites, as well as more venereal disease, tuberculosis, improper housing, poverty, and so forth.[15] Children of south European immigrants are more likely to get into jail than are those of native stock.[16] North

[14] For an appreciation of the importance of these limitations on the development of American Indian culture, see Radin, Paul, *The Story of the American Indian.* New York: Boni and Liveright, 1927. Pp. xiv + 371; and Wissler, Clark, *The American Indian, An Introduction to the Anthropology of the New World.* New York: Oxford University Press, American Branch, 1922. Pp. xxi + 474.

[15] For illustration of this point, consider the study of Negro neighborhoods, housing, schools, and recreation in sixteen cities made under the direction of Dr. Woofter. It is a type of statistical study which can be of invaluable aid in understanding the present status of the American Negro, but at the same time is of no real importance in the ultimate explanation of the relative significance of the biological and cultural factors in race relations. Woofter, T. J., Jr., *Negro Problems in Cities.* New York: Doubleday, Doran & Co., 1928. Pp. xiii + 284.

[16] For two excellent studies of the present status of European immigrants in the United States which, like Woofter's study, also fail to tell us anything about the inborn racial factors involved, although both have great practical value, see Carpenter, Niles, *Immigrants and Their Children.* Census Monographs, VII. Washington: Government Printing Office, 1927. Pp. xvi + 431; Brunner, E. deS., *Immigrant Farmers and Their Children.* New York: Doubleday, Doran & Co., 1929. Pp. xvii + 277.

American Indians have not been notably successful in business and the professions. Mexican immigrants constitute a grave problem for social agencies to deal with in southwestern cities of the United States. Such random samples may be selected *ad libitum*. Under careful scrutiny, they furnish real data for the understanding of race relations.

Much valuable statistical work has been done in this phase of cultural analysis. There has been little attempt to compare widely differentiated cultures; the effort has been rather to describe specific aspects of some one general culture system which vary with the racial identity of the sub-groups in a larger population unit. The merits of these studies have been in the two directions which the layman refers to as the practical and the theoretical. Practically, they have furnished the basis for needed social reform. Theoretically, they have laid the groundwork for improvement in our racial speculations. They are dangerous only when used in attempts to fix the cause of maladjustment in either race or social status.

This danger has not been entirely avoided. Over-emphasis on the use of statistical units based on fictitious racial identity of subjects when, as a matter of fact, the selection of cases has been actually dependent on group visibility, has marred much of this work. As a random illustration of this likely error may be mentioned a proposed study of Indians in a large eastern city. In a laudable desire to assist the painless assimilation of our Indian population, a social agency planned to study all Indians, said to be about one hundred in number, residing in a certain urban area. The Indians in mind were persons who had one-sixteenth or more Indian blood, to their own personal knowledge, and who could be located because their assimilation into the larger population was imperfect or incomplete. The undoubted and uncounted hundreds of persons of similar Indian blood who had been swallowed up by the white and Negro population were not thought of as Indians because their native origin created for them no peculiar problems. Yet their importance in such a study is beyond question since they are the very ones who have undergone the process of assimilation. If it is visibility which we wish to use as the criteria for selecting cases, let us say so, and make no further claims. Meanwhile, it should not be forgotten that visibility and race are two quite different things.

3. The determination of psychological characteristics may be attempted by the comparative study of anatomical structures and physiological functioning, and by the application of a great variety of what are loosely known as "psychological tests."

Strictly speaking, the social scientist has little business bothering with either of these methods, beyond accepting the findings of the biological scientists and making such applications to questions of race relations as may be legitimate. This point of view is not likely to be seriously challenged in so far as it applies to the field of anthropometry. There are many, however, who will no doubt claim that psychological testing of races is so dependent upon a thorough understanding of the social factors involved that if the measurements are to be made by psychologists without a broader training in the whole field of race relations, their results may be used only after painstaking examination by properly equipped social scientists.

Attempts to measure the cultural potentialities of racial groups by means of mental tests differ not at all in principle from similar attempts in the field of comparative culture.[17] A comparison of the cultural achievements of two races is theoretically based upon the same procedure as the comparison of their mental scores. The fact that the second method is experimental and the first historical means, of course, that there may be greater control and more impressive use of charts and tables in the second than in the first. The validity of both depends on the same unproved assumption.

This assumption, permitted to drift into the background by mental testers and rarely thought of by others, is that, given the same task to accomplish, two groups of individuals will produce different degrees of success in direct proportion to their innate ability to accomplish such tasks, provided environmental influences are, and have been, identical.

The joker lies in the provision that environmental influences must be identical if the results are to be interpreted in terms of race. No psychologist has thus far been able to fulfill this provision beyond reasonable doubt. The study which probably comes nearer than any other of similar extent to meeting this necessary condition is that by Peterson and Lanier on the comparative abilities of whites and Negroes.[18] Kirkpatrick's analysis of immigrant in-

[17] Mental tests have been most widely and effectively used in comparing Negroes and whites. For discussion of their results in this field and their value in race relations, see Young, Donald, Editor, "The American Negro," *The Annals of the American Academy of Political and Social Science*, CXXXX, No. 229, November, 1928. The articles in this issue which deal specifically with this subject are: Viteles, M. S., "The Mental Status of the Negro," pp. 166–78; Peterson, Joseph, "Methods of Investigating Comparative Abilities in Races," pp. 178–86; and Thompson, C. H., "The Educational Achievements of Negro Children," pp. 193–209.

[18] Peterson, Joseph, and Lanier, Lyle H., *Studies in the Comparative*

telligence should also be mentioned here as one which reduces the importance of differential environments to a minimum.[19] Yet even these authors, more careful than many, definitely refuse to claim the complete elimination of outside influences on the colored, foreign, and native white subjects tested. They do claim to have minimized such factors, but no one can tell the extent to which the residue was responsible for the differential scores obtained.[20]

After all, such tests are cultural in nature and require the subjects to express themselves through cultural media. There is no special magic in triangles, squares, circles, numbers, animal pictures, and similar simple devices, which will automatically eliminate the influence of formal and informal education, varying incentive, malnutrition, and the like. No

Abilities of Whites and Negroes. Mental Measurement Monographs, Serial No. 5. Baltimore: The Williams and Wilkins Co., 1929. Pp. iii + 156. The concluding sentences of this monograph are significant: "What the net effects of these differences in samplings (favoring the Negroes in the comparison of college students) and in school training (favoring the whites) really are on the test scores, we cannot say without further investigation. It is best not to speculate on them here." (P. 152.)

[19] Kirkpatrick, Clifford, *Intelligence and Immigration.* Baltimore: Williams and Wilkins Co., 1925. Pp. 127.

[20] Professor East in a chapter on racial traits in his *Heredity and Human Affairs* discounts the criticisms of psychological tests as measures of innate ability, in so far as they are based on the belief that cultural factors cannot be eliminated. "One cannot measure intelligence except by observing its results after development. But this is no reason for assuming that the observations made cannot be interpreted in terms of native intelligence. . . . People of a different culture are undoubtedly handicapped when undergoing the current psychological examinations, but it is doubtful whether the handicap is significant." (P. 177.) Professor East, a biologist, is cited here to make clear a point of view in opposition to the author's, although it is doubtful whether any large number of leading psychologists would agree with Dr. East in totality. Cf. East, E. M., *Heredity and Human Affairs.* New York: Charles Scribner's Sons, 1927. Pp. vii + 325.

one really claims there is, except the pseudo-scientists and laymen who use the work of others as it was never intended to be used.

Another point to be considered is that artificial tasks administered under experimental conditions are not necessarily indicators of general or specific abilities essential for competition in modern civilization. Boys who are dropped from college may make the best of business men. French Canadians, Chinese, South Europeans, and others, who score low on another man's test are not thereby proved unfit to till the soil, build great cities, or create a lasting art.

The inference should not be drawn, however, that such experimental work is thought to be without real value. The only suggestion here made is that the results should be used in their proper field of application. Army mental tests were useful during the World War for practical army purposes, but not in the post-war period for biological interpretations. Schools and colleges may use mental tests for admissions and other administrative purposes, but racial theorists should be most cautious in utilizing such materials as the basis for prophesying racial futures. Industrial plants may profitably encourage psychological measurements as an aid in personnel selection, but the further interpretation of the results obtained in attempts to make them apply to race relations as such requires qualification. *It is their misuse rather than their use which causes trouble.*

In conclusion we can only say that wherever the student of race relations is concerned with the present status of a racial group, mental tests as well as cultural achievement may properly be used as aids in the solution of practical problems of social adjustment. If French Canadian children show a low I.Q., this fact must be taken into consideration in planning their education and careers. If Negro school children do not test as high as do white children, the consequences of their retardation are not wiped out by explaining that it may all be due to social handicaps. Race relations are unquestionably affected by the mental and social development of a group as indicated by psychological scores and studies of social status. The exact program for improvement, however, will be directly affected by the degree to which racial or environmental explanations of existing differentials are accepted.

Wherever the interest lies in limitations placed on group relationships by innate racial qualities, it is not out of place to issue a warning that what has been done in the past or what is being accomplished now is not necessarily proof of the upper limit of future success. The statistician who neglects this warning and assumes that present correlations are fixed for years to come may easily be misled.

If a hesitant prophecy may be here permitted, the answer to the problem of innate racial qualities will probably be found through the future work of the biological, and not the social, scientists. But even this timid belief is based on a multitude of unsafe assumptions!

VI

STATISTICAL STUDIES OF CRIME AND THE ADMINISTRATION OF JUSTICE

C. E. GEHLKE

BY THE term "criminal statistics" is meant in this paper the statistical manipulation of data about crime and criminals. The data themselves are only incidentally interesting to us. The field of criminal data in which statistical method is used may be roughly divided into the following classes, chosen largely for their practical convenience: (*a*) crimes, their number, and distribution in time and space; (*b*) the causation of crimes; and (*c*) the treatment of criminals by police, courts, institutions, probation, etc. At the present time we are greatly hampered in the United States by the scarcity of accurate data in each of these fields. In the matter of prison statistics we have a good beginning in the work of the Federal Census Bureau. Otherwise, we are dependent upon scattered local data, lacking in large degree any sort of uniformity the country over.

As a result of this serious defect, nearly all of the recent statistical studies have been made upon the basis of data gathered *ad hoc* at great expense and with much labor. In practically every such case the material was of a nature that might reasonably be expected from a good system of official statistics. For example, the statistical data of the surveys of criminal justice in New York, Illinois, Pennsylvania, Missouri, Georgia, Cleveland, Cincinnati, had to be gathered by the laborious process of summarizing individual cases from the official records of police, courts, and institutions.

Methods in any sort of statistical work are determined by two sets of limiting conditions: the purpose of the study, and the nature of the data. The purpose may be administrative or scientific. (We may ignore the propagandist type of purpose, although there are plenty of examples.) There is naturally no sharp line of demarcation between the administrative and the scientific purposes. Some administrators have a scientific point of view and scholarly capacity.

The relation of method to purpose will be discussed here in only the most general terms. Obviously, without a careful analysis of individual studies the elaboration of the topic is impossible. Ultimately the method must be chosen with the reader in mind. The expert can understand complicated methods; an inexpert reader demands simplicity. Official statistics published for the enlightenment of the general public call for the simplest and the most ingenuous technique. The man in the street has a rather suspicious attitude toward figures of any sort, particularly if he believes there is some ulterior motive behind them. There is also the methodological argument for simplicity. The more complicated the method, the more numerous are the logical and mathematical assumptions underlying it. That these assumptions are in harmony with the nature of the data is sometimes difficult to establish. These points are merely in the nature of cautions, not prohibitions.

The second general limitation, that presented by the nature of the data in this field, has two principal forms. The first is that here, as everywhere in the social sciences, we must use for technical purposes terms which for ages have had a common usage. The second is that most of the categories into which the data fall are qualitative and not quantitative.

1. The simplest illustration of the first form is to be found in the word "crime." This word includes within its hospitable boundaries a very wide variety of meanings, both substantive and theoretical. What is "crime" in one age and land is a mere fault in manners in another. The diversity of general definitions of the term indicates that even the universal elements of the definitions cannot be agreed upon. A second difficulty is one peculiar to countries having a federal system of government like that of the United States. Each of our forty-eight states has jurisdiction over all offenses except those which are specific violations of federal law and not under a joint jurisdiction. That means that there are differences in the substantive law, and in the administrative and procedural law, which governs the treatment of the criminal. Concretely, "burglary" is an offense the legal definition of which varies from state to state, although, of course, only in details. In many states the prosecuting attorney may enter a motion of "nolle prosequi"; but in New York that has been removed from criminal procedure. There the judge may act on a motion of the prosecutor to dismiss the case. Some states have prosecution on "information," others only on "indictment," and still others have both possibilities.

To illustrate the practical difficulties which this situation presents, attention may be called to the procedures which had to be adopted in order to secure a workable uniformity in the reporting of crimes by the police. The International Association of Chiefs of Police, through its Committee on Uniform Crime Records, has analyzed the statutes of each state with respect to the definitions of the more frequent and serious felonies. For each state there has been prepared a list of the specific statutory offenses which are, for purposes of reporting, to be included under the heading of, e.g., "burglary." Obviously the term "burglary" is not a perfectly homogeneous classification, but it will have to serve the practical purposes of a system of uniform crime statistics. Another example is the use of the term "probation." In one state this procedure involves very careful selection of probationers, with intensive supervision while on probation; in another state it will mean hardly more than a suspension of sentence. Obviously, to take figures of "probation" for the whole country as homogeneous data would be a statistical error.

The ultimate source of statistical data in the field of crime is almost always some kind of public record, of the police, of the courts, of penal institutions. In addition to the intrinsic variability of the data as already indicated must be cited the too often inaccurate and occasionally actually falsified records from which the facts must be transcribed.

2. The second principal weakness in the data, namely, the predominance of qualitative over quantitative categories, is easily illustrated. The classes most commonly found in tables of criminal statistics are those of offenses, dispositions, sex, location, occupations of the accused, etc. A few quantitative classes will appear, such as age, length of sentence, length of time served, amounts of fines, time consumed by the proceedings, number of previous imprisonments or convictions; and, in more recent and highly specialized studies, intelligence quotient,

various educational indexes, and so on. While apparently the groups resulting from the two kinds of classification are about equally numerous, the amount of material in the second (quantitative) is far less than in the qualitative. As time goes on and as technique is improved, we shall probably have an increasing amount of material which can be summarized and analyzed quantitatively. But this probability is still in the more remote future. From the point of view of method, the present situation means that any extensive use of statistical procedures based upon frequency tables is unavailable. What we do have available is the use of simple tables, tables of percentage distribution, and comparisons based on these percentages. Such material, of course, lends itself to the application of probability methods. But too frequently the classes are so small that this is impossible.

Turning now from the questions raised by the limitations of purpose and data, let us consider briefly to what extent we may rely upon the statistical method alone in discovering those uniformities which are sometimes called "laws." This is another way of asking whether we may attack the problems of causation in the field of crime by means of statistical measures, and how far these methods will take us.

The types of questions which arise here may be illustrated by such as these: To what extent is crime due to mental defect or disease? What is the importance of the factor of home environment? What is the influence of the infliction of punishment (upon some criminals) in deterring others from similar lawless deeds? Is a given police department functioning effectively? Do the courts of a certain city secure the maximum output of accurate determination of guilt and of the proper treatment of those found guilty? These questions are obviously not posed in a scientifically qualified manner, but they represent the raw material of which scientific questions may be made.

The first step in the scientific analysis of any set of phenomena is accurate description of the objective data. Nearly all of the study of crime, criminals, and the treatment of criminals has been, up to a very recent time, based upon observation, experience, and the kind of shrewd summarization which intelligent people can often make without any sort of statistical procedure. It is the first function of statistics to transform such observation and the result of such experience into a quantitative form by the basically simple process of counting. A veteran judge may state that it is his opinion that a large number of accused persons are never punished for the offenses they are alleged to have committed, and we should probably respect that opinion. But it is not material on which to base scientific conclusions.

On the other hand, suppose we follow each case in which a felony arrest was made in the city of Chicago in 1926 [1] (12,543 in all) and find that of this number there were eliminated without punishment: 49 percent in the preliminary hearing; 11 percent in the grand jury; 20 percent in the trial court; 5 percent after guilt was established; and that 15 percent were punished. We now have a statistical sample of sufficient size to conclude that the above percentages present for the given year a fairly accurate picture of what happens in this jurisdiction. But can we stop with these results, or are further meanings to be sought from the data?

[1] *Illinois Crime Survey*, p. 35.

First, we want to know whether the conditions pictured are peculiar to Chicago. We may therefore compare Chicago with New York and with Milwaukee, or with any of the other cities for which similar data are available. Further studies of Chicago cases of earlier and later date might reveal whether or not this was a result peculiar to the year 1926. If differences are revealed by any of these comparisons, then we have at once to explain what may be the reasons for these differences. The next step will be to subdivide still further the disposition groups (e.g., those eliminated in the preliminary hearing). We will ask what percentage of this sub-class was "nol prossed," what percentage was discharged, and so on. Subdivision will reveal that part of these dispositions is the work of the judge; part is the work of the prosecutor; and part, of the jury. We can also divide the cases according to the offense charged. But ultimately we shall get to the point at which further subdivision results in classes too small for reliability or at which there is no factual basis in the records on which further division can be made.

When this point is reached we have, for the time being at least, to turn to the non-statistical data. For example, in a proportion of cases, prosecutors will accept pleas of guilty to charges less serious than the original charges. If we find that one such official is accepting more of such pleas than another, we cannot hope to find the reason for the differential within the system of our statistics. It lies somewhere outside the records of the courts. It may be sought in the political situations of the two cities. It is possibly due to peculiarities of the particular prosecutors, which must still be explained. To find the reason we might have to engage in research which would ultimately involve the history of Anglo-Saxon jurisprudence from King Alfred to State's Attorney Crowe of Cook County, Illinois. Even if we had full data for every city in the United States over a period of fifty years, we should still somewhere have to fall back upon the method which may be called, broadly, the historical method. As soon as we get away from mass phenomena, our statistical procedures fail us. It cannot be too strongly emphasized, however, that in a problem such as the one just used as an illustration, statistical method is indispensable in certain stages. It affords the only means of accurate description where masses of facts of even a roughly identical type are to be dealt with.

The same general course of reasoning applies in other fields of criminal investigation. We may subject thousands of prisoners to mental tests, such as the Army tests, or to anthropometric examinations, as did Goring. Whatever the results compared with those from non-prison populations, we shall never get through statistics alone a thorough understanding of the causation of crime, unless we find some one trait possessed by all prisoners and by no persons outside of prisons. If we find that some feebleminded persons are criminals and that some are not, we evidently have to look further for the cause or causes. It is true that our procedure may reveal valuable clues. But these must ultimately be followed, not merely down the statistical paths that lie open, but down other roads as well. Here the "historical method" appears with a new name: "case analysis." Another type of study in this field, as yet but slightly demonstrated, is the time correlation of crime and other social phenomena. But here again, after we find a correlation coefficient, we have only a more accurate description of a social situation. The question still must be answered, Why is there a correlation (positive or negative, as the case may be) between phenomenon A (crime) and phenomenon B (bank clearings, let us say)?

There are, however, practical uses of statistics in which the ultimate scientific goals do not need to disturb us with inexorable questioning. As an example I suggest Professor Burgess' study of the factors which seem to influence success on parole. He has worked out a probability table on the basis of which it is possible to estimate the probable success of a parolé from three Illinois penal institutions.[2] The method is briefly this: prisoners having certain characteristics fail on parole less frequently than do all persons paroled upon the average. The larger the number of such characteristics a prisoner has, the more likely is his parole to be successful. On the basis of the number of "success characteristics" it is therefore possible to estimate his chances of success. In a situation in which the exact determination of this probability in any given case is impossible, reliance may be put for practical purposes on the prognostication offered by the table. After all, this is merely applying in another field the principles which have long been the sound financial basis of life and other kinds of insurance. Such methods may be worked out for administrative purposes in a variety of situations of a similar nature.

In conclusion, we may briefly summarize the fore-

[2] *Illinois Crime Survey*, pp. 516–40. Also, Burgess, E. W., "Is Prediction Feasible in Social Work?" etc., *Social Forces*, VII, No. 4, June, 1929, pp. 533–45, together with discussion by Rice, Stuart A., "Some Inherent Difficulties in the Method of Prediction by Classification," *ibid.*, pp. 554–58. Also, Burgess, Robert W., "The Rôle of Sub-Classification in Statistical Analysis," *ibid.*, pp. 558–60.

going rough analysis of the rôle of statistics in the field of crime. It is apparent that statistical method is called for here, as everywhere, when there is a mass of facts which can be counted. There is no part of this field in which this situation does not arise in greater or less degree. But there are serious limitations on the use of statistics through the nature of the data, through the purposes of the investigation, and, most important, through the fact that statistics must be supplemented by the historical method.

VII

THE BEGINNINGS OF JUDICIAL STATISTICS

L. C. MARSHALL

I. THE PRESENT STATUS OF JUDICIAL STATISTICS

A STUDY of the literature of judicial statistics in this country, including the systems in actual use, leads to but one conclusion: we have made only a beginning of serious work in this field.

Partly because of the prevailing attitude that crimes are a matter of social concern (whereas civil litigation is too commonly thought of as being primarily the private affair of the contending parties) and partly because of the spectacular features and consequences connected with some types of crime, the main development has been in the field of criminal statistics—using this term to mean the statistics of both crimes and criminals; and to embrace police statistics, judicial criminal statistics, and penal statistics.

In the field of police statistics, a long history of fragmentary, inaccurate, non-comparable data secured in a few scattered jurisdictions has just been closed by two generations of agitation respecting the need of comprehensive, reliable, comparable data. This agitation has borne excellent fruit in the recent publication of *Uniform Crime Reporting*, "a complete manual for police," and *A Guide for Preparing Annual Police Reports*, by the Committee on Uniform Crime Records of the International Association of Chiefs of Police. This committee, aided by a competent advisory committee and by an equally competent research staff, has provided "a complete work program and manual for the guidance of police departments and of state and national statistical agencies." In its own field, this is an epoch-making event; there is now available for installation and practical use an adequate system (of course, ever to be improved) of police statistics, including the necessary forms. records, and reports. In the related fields of judicial criminal statistics and penal statistics, this manual and this guide will exert no small influence. This will be especially true because of its uniform classification of offenses—the *sine qua non* of really constructive work.

In the field of penal statistics a great deal of constructive work has been done by individual students, by state institutions and departments, and by the Bureau of the Census of the National Government. Particularly worthy of mention is the publication in 1927 by the Bureau of the Census of *Instructions for Compiling Criminal Statistics*, "a manual for the use of penal institutions, police departments, courts, prosecutors, and parole and probation agencies." As the subtitle indicates, this manual deals not only with penal statistics, but also with judicial criminal statistics and police statistics. Noteworthy, too, is the adoption by the Bureau of the policy of providing annual censuses (instead of the earlier system of occasional censuses) of state and federal prisons and reformatories.[1] It is true that the penal data com-

[1] Leon E. Truesdell in "The Problem of Collecting and Standardizing Statistics of Crime in Forty-eight Sovereign States," *Proceedings of the American Statistical Association*, March, 1928, writes thus: "Statistics of some sort relating to crime or criminals have been collected by the Bureau of the Census in connection with each decennial census since 1850. In 1880 inquiries were planned to cover not only prison population and commitments but court proceedings and police records

piled by the Bureau of the Census throw little or no light upon many of the graver problems in the criminological field, and it is true that the same remark may be made even of the tabulations to be found in such excellent reports as *The Annual Report of the Commissioner of Correction* of the state of Massachusetts; but, none the less, penal statistics are much the most developed branch of our criminal statistics. This is not surprising. The penal data, of the kinds now being gathered, are much more readily accessible and much more simple and definite than are most of the other forms of data in the criminal field.

Between these two fairly well charted, but by no means well exploited, fields—one having to do primarily with the earlier stages of the handling of

as well, though only the prison population data and incomplete results of the police inquiry were ever published. From 1850 to 1890, therefore, the available statistics include little more than the prison population, and even these figures involve considerable variations in definition. For the year 1904 there was taken what might be termed the first of the special censuses of penal institutions, covering not only prison population but commitments and discharges as well, all classified in considerable detail. Similar and comparable figures are available for 1910 and for 1923. These three inquiries covered state and Federal prisons and reformatories, county and municipal jails and workhouses, and juvenile reformatories.

The Census Bureau is now engaged in the first of a series of proposed annual censuses of state and Federal prisons and reformatories. . . . Before the Census Bureau can well make any material expansion in its present program of annual statistics of crime, which is now limited to prisoners in state penal institutions and their commitments and discharges, it will be necessary to have a specific authorization from Congress. The present limited inquiry is being undertaken, under a general law, as a sort of experiment. . . .

The first thing, perhaps, would be to expand the scope of the annual institutional inquiry to include jails and workhouses as well as prisons and reformatories, putting it on a par with the so-called decennial inquiries of the immediate past, with the addition, perhaps, of data on paroles from these institutions. That would be simply to round out one group of criminal statistics, however, and would not be enough."

criminals and the other having to do primarily with the final stages—there is the field of judicial criminal statistics, in the main a *terra incognita*. It is true that in a few states a fair amount of statistical tabulation of the work of the criminal courts is published; it is true that a few judicial councils are beginning to cultivate the field; it is true that in a few cities voluntary crime commissions assemble considerable amounts of data; it is true that such occasional surveys as *Criminal Justice in Cleveland*, *The Missouri Crime Survey*, *The Illinois Crime Survey*, and *Crime and the Georgia Courts—a Statistical Analysis*, have thrown much light upon the operations of our criminal courts. But it is also true that present achievements in this field are as a few scattered oases of information [2] in a Sahara of ignorance with respect to the grist being ground by the judicial machine.

Bad as is the case of judicial criminal statistics, the case of judicial civil statistics is distinctly worse. In a few jurisdictions, statistics of divorce cases are available, but the data are typically dull, formal, and quite obviously more significant in what they conceal than in what they reveal. So also in a few jurisdictions (for examples, see *Ohio Statistics;* the account of the business of the Suffolk County Probate Court in the *Fourth Report of the Judicial Council of Massachusetts;* and the "Summary, Probate Courts" in the *Second Annual Report of the Judicial Council of Kansas*) the probate work is analyzed. But as for the usual run of "business cases," there is little information available, and this little is seldom presented in terms of any classification of types of cases. Rather it is

[2] Other oases of information may be found in the reports of a few municipal courts and in *Juvenile Court Statistics* (Bureau Publication No. 159) published by the Children's Bureau of the United States Department of Labor. See also Children's Bureau Publication No. 195.

typically a simple, almost crude, picture of the amount of such business coming before the courts, the amount disposed of, and the amount still pending.

If the question were to be asked: Is there a single state in which, either as a result of an individual survey or as a result of an on-going system of judicial records and reports, one could get access to a comprehensive, balanced account of the work of the judicial system of the state—covering all courts from the highest to the lowest and covering all types of cases coming before these courts—the answer is simply made. No state has assembled such data; no state has even approximated performance of such a task; and in only a few quarters is the importance of such data appreciated. This is the situation as to mere tabulations of court data. If one looks for far-reaching interpretation of data, and especially if one looks for illuminating correlations of judicial data with other types of data, he finds a void that is absolute and of Stygian darkness. In these respects we are in a pre-statistical era.

The situation is, however, far from hopeless. There are quite definite forces making for a future which promises to be bright—certainly not as dismal as the present.

It was said above that in only a few quarters is the importance of judicial statistics appreciated. Fortunately, those are influential quarters, possessed of modest powers to bring about improvement. Students of the administration of justice [3] have long appreciated the importance of the formulation and

[3] Some of the more forceful and penetrating periodical literature on the relation of statistics to judicial administration is to be found in the files of the *Journal of the American Judicature Society* and the *Journal of the American Institute of Criminal Law and Criminology*.

installation of a competent system of judicial records and reports (*a*) as a means of providing exact information which may be utilized to secure efficient operation of the existing machine by its responsible operators, and (*b*) as a means of providing that continuing flow of data so important for the guidance of judges, legislators, students, and others concerned with the progressive betterment of the machine. In this student group are to be found not only representatives of legal scholarship in our universities, but also a sizable handful of our best judges and practitioners. These, and others like them, have in various ways stimulated the movement for competent judicial statistics; and, in particular, they have stimulated the development of judicial councils, which now exist in one form or another in almost a score of our states. Among other statements which may properly be made concerning these judicial councils, it may with emphasis be said that they represent a tendency toward a more orderly and coherent organization of our court system; and, of course, this inevitably points to the importance of a more orderly and coherent system of judicial records and reports—judicial statistics.[4]

The judicial council movement then, by the very nature of the case, is a movement which in some real sense stands sponsor for judicial statistics—a fact that was well recognized in the discussions which took place at the recent organization of the Conference of Judicial Councils. The judicial council movement, however, does more; it puts in the hands of many of the councils

[4] This statement should not be understood to imply that the movement for judicial statistics is a movement solely connected with the development of judicial councils. That is not at all the case. It does seem probable, however, that these councils may take a position of leadership in a fairly rapid development of judicial statistics.

both the power and the obligation to gather data and make reports. The Ohio judicial council is a case in point. The enabling statute of the first judicial council to be definitely constituted states that the council is authorized to make a "continuous study of the organization, rules, and methods of procedure and practice of the judicial system of the state of Ohio, the work accomplished and the results produced by that system and its various parts" and to report to the legislature its "recommendations for modification." The clerks and other officers of the various courts must make to the council "such reports on such matters and in such form periodically or from time to time as the council may prescribe." Now this language does not in explicit terms set up a bureau of judicial statistics, and it does not prejudge the issue whether a bureau of judicial statistics (if ever set up) should be located in the judicial system or in some other part of the state government. But there can be no question that the powers granted the council are sufficient to enable it to force the installation of a system of judicial statistics, if the council chooses to be active and persistent in the matter.[5]

[5] The law relating to criminal statistics in the Dominion of Canada provides an interesting contrast with our own state laws on such matters and it points to a goal toward which we might well move in providing for judicial statistics. The significant sections of the act are as follows:

> 29. The clerk of every court or tribunal administering criminal justice, or in case of there being no clerk, the judge or other functionary presiding over such court or tribunal, shall before the end of October in each year, fill up and transmit to the Dominion Statistician, for the year ending the thirtieth day of September preceding, such schedules as he receives from time to time from the Dominion Statistician relating to the criminal business transacted in such court or tribunal.
>
> 30. The warden of every penitentiary and reformatory and the sheriff of every county and district, shall, before the end of October

Not all the judicial councils have powers as broad as this. But, since many of them do have these powers, there is needed only a "will-to-do" on the part of one or more of these judicial councils (preferably several of them) and a new era in judicial statistics can be rather quickly opened.

It happens that this will-to-do has already been developed. Several of the councils are moving in the direction of the establishment of systems of judicial statistics. The following pages are an account of the beginnings of such a project in the state of Ohio. They have been written in the hope that criticisms of the methods being followed and suggestions for their improvement will be made. The field is complex and uncharted, the number of skilled workers available is very small, and the consequences of errors will be so evil that it is important to secure the best possible counsel and advice.

in each year, fill up and transmit to the Dominion Statistician, for the year ending the thirtieth day of September preceding, such schedules as he receives from time to time from the Dominion Statistician relating to the prisoners committed to the penitentiary, reformatory or jail.

31. Every person required to transmit any such schedules shal from day to day make and keep entries and records of the particulars to be comprised in such schedules.

32. Every officer required to transmit to the Minister of Finance true copies of returns made by justices of the peace, shall, before the end of October in each year, transmit to the Minister true copies of all such returns for the year ending the thirtieth day of September next preceding.

33. The Secretary of State shall, before the end of October in each year, cause to be filled up and transmitted to the Dominion Statistician such schedules for the year ending the thirtieth day of September next preceding, relative to the cases in which the prerogative of mercy has been exercised, as the Minister may prescribe.

As is well known, this law, competently administered, has resulted in a fairly comprehensive system of criminal statistics.

II. Planning Judicial Statistics for Ohio

In May, 1929, The Institute of Law of The Johns Hopkins University made to the Judicial Council of Ohio a proposal looking to a comprehensive study of judicial administration in the state. Among other things, this study was to include the formulation and installation of a system of judicial statistics for the state. The proposal of the Institute of Law was accepted by the Judicial Council; and at the summer meeting of the State Bar Association a coöperating committee was appointed by that body. On December 2nd the organization meeting was held and the work is now being carried on, under the auspices of the Judicial Council.

The difficulties of the task are, of course, enormous. There are, in the first place, serious difficulties growing out of the present character of the judicial machine. It is easy to talk of a "system" of judicial statistics, but it is not easy to visualize one which will be effective in a judicial machine which itself has a minimum of system. Each separate court tends to be a unit in itself, with its own forms and records and its own methods of handling its business; and a similar statement may be made of all the other agencies affiliated with the judicial machine. There is no tradition favorable to the collection of statistics and, of course, there is no personnel trained for the task. This is not said in criticism of the courts; in view of their history, any other state of affairs is not to be expected. The unexpected thing is that, here and there, a single court rises above the murk of its past.

There are, in the second place, the equally serious difficulties growing out of the fact that little scholarly work has been done on judicial statistics. There is among students of the problem little agreement with respect to the types of court data which it is important to seek, and even less agreement with respect to what may be called, for lack of a better term, non-court-room data. Everyone knows that we are definitely interested in securing data on the judicial process proper; data on the human element in each case; and data on social situations which are relevant to the purpose in hand—but beyond this general statement there is confusion and uncertainty. And the data—whatever they may be—are, from the statistical point of view, elusive and difficult to define. The situation could be worse, no doubt, but it is in all conscience bad enough.

There is, of course, no possibility of thinking through in advance all the problems which will arise in such a task. That would not be possible even if the field were relatively simple and already carefully explored; even if much experience had already been had in the practical operation of comprehensive systems of judicial statistics. Far less will it be possible in such a complex and intricate field; in a field little explored; and in a situation where experience must be gained as the work proceeds. And, of course, there is no possibility whatever that such a task can be performed and then stand for all time—or even for any great length of time. The very phrase "competent system of judicial statistics" must continually be redefined, and the details of any such system must change as social goals change and as social organization to accomplish those goals changes. All that can be hoped is that the first attack on such a problem will be reasonably well considered in the light of expert counsel and practical experience.

In view of the largely unexplored difficulties of the

case, it has seemed expedient to fall back upon a few well-established rules: first, there should be a reasonably clear conception of the purposes which are to be served by the statistical system; second, details of the system should be set up with full recognition not only of these purposes, but also of the character of the operations concerned and the personnel available; third, no amount of preplanning can dispense with the necessity of much use of trial and error—of cut-and-fit—in such work. These abecedarian rules of statistical systems will be discussed in turn.

1. There can be, of course, neither rhyme nor reason in a system of records and reports except in terms of the purposes of that system. It has been suggested earlier that a system of judicial statistics should be thought of, in part, as a means of providing exact information which may be utilized to secure the efficient operation of the existing machine by its responsible operators [6]—this with respect to the effective

[6] The principal purposes or objectives of statistics in business administration have been stated thus by H. S. Person in "Statistical Control, Including Costs as a Factor in Production," *The Annals of the American Academy of Political and Social Science*, LXXXV, 1919, pp. 224–28.

> 1. To ascertain inner, controlling, master facts which cannot be ascertained by casual observation of the complex mass of obvious facts which constitute the experience of the business and in which they are contained.
>
> 2. To determine standards by which to value and guide current performance and in terms of which to estimate future performance.
>
> 3. To establish series of facts which suggest tendencies or permit comparisons which suggest causal relations, or at least correlation, between series.
>
> 4. To determine laws governing industrial operations.

The following statement taken from "What a Judicial Council Can Do," *Journal of the American Judicature Society*, VII, 1924, p. 159, is illuminating:

"A good deal of the business of administering justice is just plain business. No business conducted in every city and county in the state, requiring uniformity of method and precision could possibly be done expenditure of time and money, the effective use of personnel, etc. If it is not sufficiently implied in this statement, then there should be explicit recognition of the fact that the litigants are as vitally interested in such issues as are the officials who operate the machine. So also is the general public which these officials serve. In part, too, the system should be thought of as a means of providing that continuing flow of data so necessary for the guidance of scholars, judges, legislators, and others interested in the improvement of the judicial machine—in ascertaining those points at which some type of readjustment is expedient either because this readjustment will more efficiently serve old social forms and needs or because the readjustment is necessitated by the development of some new social situation.[7]

well without a thorough system of record keeping. To find out what the courts are doing and how various laws, rules and methods are succeeding, from the present scattered records, would require months of time and a small fortune in money. A single balance sheet would be of little value. What is needed is continuing records kept in a uniform manner, and made available to the public in a convenient form."

[7] The case for judicial statistics has been put thus by Willoughby, W. F., in *Principles of Judicial Administration*, pp. 345–47:

"It is doubtful if a country has a more important task than that of the administration of its civil and criminal law. It is impossible to perform this task efficiently without full information regarding the causes to be handled and the action taken in reference to such causes. The statement is often made that the amount of litigation is steadily increasing with the result that the burden thrown upon the courts is correspondingly augmented. We do not know whether this is so or not; or, if so, the extent of the increase and the classes of cases chiefly responsible for it. We have no authentic data as to whether the business now coming before the courts is of the same character as formerly, or whether it is handled with greater or less dispatch, economy, and certainty. The statement is also often made that the country is being swept by a crime wave, or that crime generally is on the increase. We have no certain knowledge as to this. Prison statistics may give us some information regarding the successful prosecution of crime,

2. The details of the system should be worked out not only in terms of its purposes, but also in full recognition of the character of the operations concerned and the personnel available. For example, the system of records of original entry should be so drawn as to reflect the data deemed valuable and in practice available; so organized as to economize the efforts of those who make the entries and the subsequent summarizations of those entries; and so conceived as to be within the bounds of practicability in view of the resources of funds and personnel available for the purpose. Again, schemes of summarization and presentation should be worked out in such a way as to organize the data in the ways most useful to serve the ends sought. This may be interpreted to include

but they throw no light upon the more important questions of the number of offenses reported to the authorities, the number of prosecutions instigated, the action had on such prosecutions, etc. Again, we know little or nothing regarding the classes of the community most responsible for crime. The assertion is often made that the prevalence of crime in this country is due chiefly to our large foreign born population and to the illiterate class. We do not know whether this is a fact or not. Within recent years there has been a great development of the system of paroling prisoners. This system is receiving a constant development without any reliable information as to how it has worked in practice or the essential requirements of a successful system. It may well be that the results are good in a state where provision is made for an adequate corps of competent parole officers, and that nothing but bad results are secured in a southern state where the criminals paroled are largely negroes, where the parole officers are selected largely with reference to political considerations, or where parole officers perform their duties in a lax or perfunctory manner. Constant experimentation is taking place in the way of creating new courts and in the adoption of modifications in procedure. The states differ among themselves in respect to the use of the grand jury and in the requirement of unanimity on the part of petty juries for a decision. We have no means of measuring the results obtained under these different systems for the purpose of determining under which the best results are obtained. We have it constantly brought to our attention that there are many more murders reported in the single city of New York or Chicago than a "framework of reporting" to meet the needs of executive officers and also to include a "framework of publication" for the information of students, legislators, and the public generally. Then, too, thought should be given to securing a collecting agency or a body of personnel which, at the very least, is competent to operate the system; and which, preferably (while operating a relatively simple system) is competent to make special inquiries to answer new problems. It is axiomatic that the basic system should be limited to a relatively small number of fundamental matters; but it should be so conceived as to facilitate special inquiries and so drawn as to facilitate correlation with other types of social data.

3. In such a task there is need to use trial and error or cut-and-fit. There must be experimental "try-

in the whole of Great Britain. We lack the data to enable us to determine just why this should be so, whether responsibility rests upon the character of our population or in our defective system for the detection, prosecution, and punishment of crimes of this character.

"In the field of civil law we know that important changes have taken place in respect to the character of litigation coming before the courts. In a general way we are aware that actions for damages brought by employees against their employers have diminished in number as the result of the establishment of workmen's compensation commissions and changed methods of handling this class of cases. It is said that actions involving the title and possession of real estate have greatly diminished in number as the result of the work of title examining and guarantee companies and the gradual clearing up of questions of titles as the country has passed out of frontier to more settled conditions. On the other hand, new classes of cases, unknown to the courts of earlier days, have arisen. We have no accurate methods of measuring these changes. In the field of criminal jurisprudence, it is also likely that important changes have taken place and are now occurring. Crimes against property may have increased or decreased relatively to those against the person. There may be an increase or decrease of serious crimes as compared with those of lesser seriousness.

"All of these and scores of other points, it is of the utmost importance should be known. If we do not know the facts, it is impossible to take the proper action in reference to them."

Analysis of Civil Cases (Not Including Divorce, Annulment, Alimony)
Common Pleas Courts of Ohio, Filed April 1, 1930 to June 30, 1930

Docket No. of Case............

1, 2,....Court of Common Pleas of Cuyahoga County. 3, 4, 5,...........Trial Judge..........

Title or Style of Case

Plaintiff

vs.

Defendant

Attorneys

(List opposite plaintiff and defendant)

Special Questions

	Yes	No
6. Is this a traffic accident case involving motor vehicle? *Check*	----	----
7. Is this a traffic accident case involving train or trolley? *Check*	----	----
8. Does this case arise from the purchase of goods on instalment? *Check*	----	----

10. Origin of Case

(Check [with x] one item only)

---1. Originally filed in Common Pleas Court
---2. From justice of the peace
---3. From municipal court
---4. From probate court
---5. From insolvency court
---6. From industrial commission
---7. From other court, board, or commission

13, 14. Residence of Parties

(Check [with x] parties plaintiff on the left and parties defendant on the right; use only one check in each column)

Pltf. Dft.

---1. All resident in Ohio 1.---
---2. All non-resident in Ohio 2.---
---3. Some residents and some non-residents of state among the parties 3.---

15-18. Status of Parties

(Check [with x] parties plaintiff on the left and parties defendant on the right; use only one check in each column)

Pltf. Dft.

---01. Individual person or persons only 01.---
---02. Individual person or persons acting as administrator, executor, trustee, or receiver 02.---
---03. Private corporation (including public service corporation) 03.---
---04. Private corporation acting as administrator, executor, trustee, or receiver 04.---
05. State municipal corpo- 05.

20-23. Mode of Disposition of Case

Principal Action *(Check [with x] one item only in left hand column, when and if case is disposed of)*

Counterclaim *(In case of counterclaim, check [with x] one item only, in right hand column)*

Default, Cognovit, Consent, Confession

---01. Judgment by default 01.---
---02. Judgment by cognovit 02.---
---03. Judgment by stipulation, consent, or confession, not including cognovit 03.---

Dismissed or Settled

---10. Dismissed by plaintiff without prejudice before trial 10.---
---11. Dismissed by plaintiff without prejudice during jury trial 11.---
---12. Dismissed by plaintiff without prejudice during non-jury trial 12.---
---13. Dismissed by plaintiff with prejudice before trial 13.---
---14. Dismissed by plaintiff with prejudice during jury trial 14.---
---15. Dismissed by plaintiff with prejudice during non-jury trial 15.---
---16. Dismissed by court before trial 16.---
---17. Nonsuit or dismissed by 17.---

36-39. Cause of Action: Type of Case

Principal Action *(Check [with x], in left hand column below, cause of action. If more than one cause, check each.)*

Counterclaim *(Check [with x], in right hand column below, cause or causes of action in counterclaim.)*

Injuries to Persons or Chattels

.......01. Slander, libel 01.......
.......02. Malicious prosecution; malicious abuse of process 02.......
.......03. Malpractice 03.......
.......04. False arrest; false imprisonment 04.......
.......05. Assault, battery; other bodily injury not resulting in death 05.......
.......06. Wrongful death action or other action arising from injury resulting in death 06.......
.......07. Action for loss of services (of wife; child; employee) 07.......
.......08. Replevin 08.......
.......09. Damages for converting, taking, detaining, or injuring chattels only 09.......
.......10. Injury to person or chattels, not otherwise covered 10.......

Land Actions

.......20. Action to recover land, including forcible entry and detainer 20.......
.......21. Specific performance or other action arising from sale of land (excluding foreclosure) 21.......
.......22. Cancellation, rescission, or reformation of deed, lease, mortgage or contract relating to land (fraud, mistake, etc.) 22.......
.......23. Foreclosure (with or without action on note) 23.......
.......24. Action between landlord and tenant (excluding action under 20 above) 24.......
.......25. Action arising from oil, gas, mining, etc. rights 25.......
.......26. Trespass, waste, nuisance, negligent or other injury to land 26.......
.......27. Quiet title 27.......
.......28. Partition 28.......
.......29. Other land action 29.......

corporation)

___6. Jurisdiction acquired without service by waiver, acknowledgement or appearance

___7. No service had and no jurisdiction acquired — one or more defendants

12. Special Relief

(Check [with x] every item which applies in this case)

___1. Attachment—real property

___2. Attachment—personal property

___3. Garnishment

___4. Arrest before judgment

___5. Injunction granted

___6. Injunction denied

___7. Receivership granted

___8. Receivership denied

items 01 and 05 above

___11. Other combination of items 01 to 07 inclusive 11.___

24, 25. Common Pleas Judgment

Principal Action *(Check [with x] one item only in left hand column below)* — **Counterclaim** *(If counterclaim, check one item only in right hand column)*

___1. For plaintiff or petitioning party, in whole or in part, on principal action — Against defendant on counterclaim 1.___

___2. For defendant or against petitioning party on principal action — For defendant on counterclaim, in whole or in part 2.___

___3. No judgment — new trial granted 3.___

___4. No judgment — for other causes 4.___

___5. Other judgment (*Check only in error cases*) 5.___

___26. Judgment on pleadings, notwithstanding verdict 26.___

___27. Facts found by court after both parties have demanded directed verdict 27.___

Special Question, 9. If jury trial, was any jury discharged on account of disagreement? *Check:*

___1. Yes ___2. No

Non-Jury Trial

___31. Facts found by court—jury waived 31.___

___32. Facts found by court in non-jury or equity case 32.___

___33. Submitted to court on agreed statement of facts 33.___

___34. Submitted to referee or master 34.___

___35. Submitted to arbitration 35.___

Other Disposition

___41. Removed to Federal Court 41.___

___42. Otherwise closed or disposed of 42.___

union agreement; boycott; strike; picketing; etc.

........37. Breach of agency or professional service contract 37........

........38. Corporation or partnership—dissolution 38........

........39. Stockholder's suits other than for dissolution 39........

........40. Stockholders' liability 40........

........41. Lien, pledge, or storage of goods or securities (excluding 32, Foreclosure or 08, Replevin) 41........

........42. Carriage or shipping contract (freight or passenger) 42........

........43. Insurance 43........

........44. Breach of building construction contract 44........

........45. Fraud, misrepresentation, deceit—damages only 45........

........46. Cancellation, reformation, or rescission of commercial contract or instrument 46........

........47. Unfair competition or trade practices 47........

........48. Accounting 48........

........49. Other action arising from commercial transaction 49........

Special Actions

........51. Habeas corpus 51........

........52. Mandamus 52........

........53. Appropriation of private property 53........

........54. Bail bond 54........

........55. Assessment or tax suit 55........

........56. Interpleader 56........

........57. Proceedings in aid of execution or to preserve assets (after judgment in prior case) 57........

........58. Action to vacate, set aside or open prior judgment 58........

........59. Change of name 59........

Other Actions

........61. Will, or administration of estate (excluding testamentary trust) 61........

........62. Testamentary trust 62........

........63. Other trust 63........

........64. Breach of promise to marry 64........

........65. Alienation of affections 65........

........66. Guardianship or adoption 66........

........69. Miscellaneous 69........

Amounts

(Write in total amounts, as indicated below, for ALL parties plaintiff and ALL parties defendant.)

27.	Total costs taxed to plaintiff	$____________
28.	Total costs taxed to defendant	$____________
29.	Total claimed by plaintiff	$____________
30.	Total verdict for plaintiff	$____________
31.	Total judgment for plaintiff	$____________
	(If counterclaim or cross-action, write in amounts below)	
32.	Total claimed by dft. vs. pltf.	$____________
33.	Total claimed by dft. vs. other dft.	$____________
34.	Total verdict for defendant	$____________
35.	Total judgment for defendant	$____________

Dates

(Write in all dates [figures only] called for below which apply in this case)

	Mo.	Day.	Yr.	Intervals
A. Petition or transcript from lower court filed	____	____	____	
B. First demurrer or motion, prior to answer, filed	____	____	____	
C. First answer of defendant filed	____	____	____	
D. Last pleading filed	____	____	____	
E. Trial begun	____	____	____	
F. Trial concluded	____	____	____	
G. Motion for new trial granted or denied	____	____	____	
H. Judgment entered or case otherwise terminated in Common Pleas Court	____	____	____	

When judgment is entered, or motion for new trial is granted, or the case is otherwise disposed of, this blank should be completed. On the first of each month, all completed blanks should be sent to Miss Nettie Nulton, State House Annex, Columbus, Ohio.

____
(Name or initial of person filling out blank)

The back of the blank may be used for remarks or explanations concerning any item on the blank.

outs" of the various forms, records, and reports in selected jurisdictions and these try-outs should, of course, be in the courts themselves and in connection with the actual business of the courts. After try-outs come revisions and further try-outs. It need hardly be said that at this stage both infinite patience and infinite tact will be demanded.

So much for a necessarily incomplete sketch of the character of the study.

The practical program of action for the immediate future is thought of as centering upon three tasks: first, an analysis of the grist currently being ground by the judicial machine; second, a corresponding analysis of data from other jurisdictions; and third, the beginnings of the formulation of a system of judicial statistics. Since the third of these tasks must, for the most part, wait upon the accomplishment of the first and second, it will be passed by with but this brief mention and the foregoing sketch of the character of the work.

1. The first task is the analysis of the grist currently being ground by the judicial machine.

Beginning January 2, 1930, data sheets dealing with civil cases, other than divorce, in the common pleas courts of the state were placed in the hands of the clerks of court, along with a set of instructions to be followed in the analysis of every such case (perhaps 15,000 in all) filed between January 2nd and March 31st, inclusive. For the period from April 1st to June 30th, a modification of this data sheet (one giving more attention to procedure, providing for a more detailed classification of cases in the familial field, and correcting certain deficiencies in the original sheet) is utilized for this same type of cases. This data sheet is attached. Meanwhile, work is under way with data sheets for the appellate courts, covering perhaps 3,000 cases, and with data sheets for the supreme court covering some 1,300 cases. Beginning July 1st, corresponding work will be undertaken on the criminal cases in the common pleas courts, which will total perhaps 25,000 cases in a year; and on the divorce cases, which will amount to perhaps 20,000 cases. Analysis is also to be made of the work of probate courts, municipal courts, and other minor courts. Data will be secured, also, concerning the work of such administrative commissions as the Industrial Accidents Commission and the Public Utilities Commission. This is a program of considerable magnitude. When carried through it will reveal in detail what the judicial machine of Ohio is doing.[8]

Of course, the data sheets used in this analysis of the current grist of the machine are not thought of as being at all appropriate for use in any final system of records and reports. But equally, of course, they do serve as try-outs of definitions and as try-outs of systems of classification. As regards court data, probably the most difficult problem of definition and classification is concerned with types of action. While a good deal of literature is available concerning the classification of crimes for statistical purposes, very little is available on the classification of civil business cases, little on the classification of cases having to do

[8] The term judicial statistics must not be too narrowly conceived if a system of judicial statistics is to be of the greatest usefulness. We need to have reliable accounts not only of the courts in relation to judicial administration, but also of the executive (the prosecutor, the sheriff, the coroner, the police, *et al*) in relation to judicial administration, also of all probationary, penal and reformatory institutions, and also of certain types of administrative commissions. Fortunately, the problem seems to be one that permits attacking one post after another and in particular it seems feasible to make a frontal attack upon the citadel—court statistics—without waiting for the various outposts to fall.

with familial matters, and not much on the classification of special types of action. As regards non-court-room data, difficulties of definition and classification are accentuated by difficulties arising from the fact that there is little guidance available concerning what "social data" (*a*) *should* be secured and (*b*) in the processes of court actions *can* be secured; and yet, if the system of judicial statistics is to be something more than a sterile record of court activities, if it is to tell something of the relationship of litigation to other social phenomena, if it is to reflect something of the human impacts of law—then something must be done in this matter of securing non-court-room data. Something must be made available concerning the social and individual backgrounds of our criminal cases, our divorce cases, and our business cases. This will be most difficult. Despite the large volume of writings and comments by social scientists upon the operations of our judicial system, where in the formal literature of social science can one find even a fragmentary analysis of the non-court-room social data which *should* be secured and in the processes of litigation *can* be secured?

2. Although the analysis of the current grist of the judicial machine in Ohio will be an important preliminary step in the formulation and installation of a system of judicial statistics, it should be accompanied by a corresponding analysis of the current grist in two or three other jurisdictions. This is important not only as a check upon the work in Ohio; not only for the comparative data it will yield for detailed studies in many aspects of judicial administration; but also as a means of securing such a broad base for the formulation of state-wide systems of judicial statistics that there will be a strong probability of the results

being taken over by other states. While it is not at this time possible to give details concerning this second aspect of the immediate program, it can be said that arrangements are being made for the work to be done. Indeed, the work is already actively under way in Maryland.

Thus far there have been presented, for criticism and constructive suggestions, a statement of the way in which the general character of the study is visualized and a statement of the two outstanding aspects of the immediate program of action. There should now be added (again for criticism and constructive suggestion) a sketch of the organization which is being utilized in carrying on the work, for mistakes in this particular may be quite as costly as mistakes in the plan of study.

At the head of the work in Ohio (which is all being done through the Judicial Council) is the *Governing Board* composed of the nine members of the Judicial Council, eight members of the coöperating committee of the State Bar Association, and two members of The Institute of Law. A *Committee of Direction* of this Governing Board, made up of the Chief Justice, the Chairman of the Bar Association committee, and the Institute representatives, has direct charge of the work.

An *Advisory Committee on Judicial Statistics* is being set up with the following elements represented: the committee of direction; experts in the formulation of records, reports, and systems of statistics; persons having practical daily contact with making out such records and with using them; officials having strategic positions in the matter of systems of statistics; and members of other judicial councils who have displayed an especial interest in judicial statistics.

In connection with this general advisory committee there are associated several *Committees on Classification;* a committee on the classification of civil cases, one on criminal cases, one on cases in the familial field, and one on non-court-room data in certain fields. Working with these classification committees (whose members can be expected to devote but a fraction of their time to the work) is a small *staff* of full-time persons.

In the analysis now being made of the current grist of the judicial machine, there is a *field supervisor*, a practicing attorney with a long record of intimate contact with court processes. The *field workers* are the clerks of court who, in accordance with the law, are being asked by the Judicial Council to perform this work. The Committee of Direction had some hesitation in having this field work done by persons typically unfamiliar with statistical work. It was recognized, however, that the clerks do have other backgrounds which are highly important; that in the larger jurisdictions special assistants having legal training could be provided, and this has been done; and especially it was recognized that it is of great strategic importance to have those perform the preliminary stages of the work who will presumably later be key persons in the operation of a system of judicial statistics. It is as yet too early to know whether the advantages of using the clerks of courts as the field workers of the study are to outweigh the disadvantages; the necessary checking of the data has not yet been made. All that can now be said is that up to the present time the advantages seem to have been greater than was anticipated and the disadvantages seem to have been less than was anticipated.[9]

It goes without saying that the Institute of Law must provide statistical and clerical aid, facilities for processing the data sheets which come in, contacts with expert counsel and advice, and all the other detailed facilities demanded by such an enterprise. In particular, the Institute must handle the statistical tabulations for the states in which it is working and with which it is coöperating. Only by detailed handling of the data for a period of years can difficulties be ironed out and a uniform system that will really work be constructed. During the period that the Institute thus serves as a clearing house it may have to edit, print and distribute forms, tally sheets, manuals of instruction, etc. Since it will also need to give continual aid and advice to the reporting units as to the proper utilization of the recording and reporting system, perhaps it should contemplate the preparation of periodical bulletins or releases containing running comment and suggestions on judicial statistics. Perhaps it will need to maintain analyses of current laws and decisions which in any given jurisdiction tend to modify the scheme of classification. Such activities are mentioned, not as elements of a fixed program, but as examples of problems which will arise and must be met as the work proceeds.

The positions taken and implied in the foregoing pages may be summarized thus:

1. While it is highly important to have competent systems of judicial statistics, no one would be so

[9] Grateful acknowledgment should be made of the interest and coöperation of the bench and the bar. In some jurisdictions, members of the local bar associations have been of great assistance in getting the work properly started, while judges have by rule of court greatly facilitated securing data from attorneys.

foolish as to suppose that the formulation and installation of a system of judicial statistics is, in and of itself, a sufficient goal. In some real sense, such work is but preliminary to the really vital tasks—those of operating the judicial machine effectively and those of analyzing the human effects of law and litigation. As regards this latter task, it is self-evident that if judicial statistics are to be of substantial service, the judicial data must be gathered and arranged in relationship with other types of social data.

2. In this matter of judicial statistics we are in the pre-statistical stage. Far from being able to use advanced statistical techniques in the study of judicial data, we have barely begun to think about gathering reliable data. A modest beginning has been made in criminal statistics, but even in this field the judicial area has received less attention than the police and penal areas.

3. There is reason to believe that the time is ripe for a definite advance in securing a flow of reliable comparable data in the judicial field. For a combination of reasons, it seems best to regard the state as the unit of work, but it is to be hoped that state systems of judicial statistics will be so formulated as at once to yield comparable data and ultimately to become parts of a larger system, preferably a national system.

4. The difficulties to be surmounted in such a task are almost overwhelming. They are certainly so great that it were foolish to talk of "solving" them in the immediate future; but that is no reason why they may not at once be attacked—attacked somewhat comprehensively and with reasonable hope of ultimate success.

5. A definite project is under way in Ohio, and also in Maryland, looking toward the formulation and installation of a state system of judicial statistics. The salient features of this project have been stated in the hope that criticisms and suggestions for improvement may aid this project in measuring up reasonably well to its opportunities.

VIII

PROHIBITION: STATISTICAL STUDIES OF ENFORCEMENT AND SOCIAL EFFECTS

JOHN C. GEBHART

WE NEED a scientific approach to the study of the liquor problem in this country. Too often protagonists, either for or against prohibition, have presented only those facts which support their position, and have ignored or failed to give proper weight to those facts which are unfavorable. The result is a distorted and confused picture. We may well follow Professor Chaddock's counsel: "When several factors are involved in producing a specific result, conclusions should not be drawn from the measurement of only one factor. The attitude of mind should not be that of the debater who counts on stating his case in the strongest possible terms, allowing his opponent to check up and refute by such arguments as he can find. The effort of the scientific investigator should be to weigh and measure every known factor in the problem before hazarding a conclusion."

It must be admitted that the known factors in our problem are relatively few. Before prohibition, it was possible to gauge accurately the annual per capita consumption of liquor. Now that the liquor traffic has been outlawed, we must rely on various indexes of intemperance and on federal enforcement statistics for indications of the amount of liquor available and the continuation of intemperance since prohibition became the law of the land.

Certain indexes of intemperance are accepted by both Wet and Dry partisans as valid: deaths from alcoholism, alcoholic insanity, acute alcoholic patients treated in public hospitals, and arrests for drunkenness. Unfortunately, records of alcoholic insanity and acute alcoholism are available in only a few states and cities. We must, therefore, rely chiefly on deaths from alcoholism and arrests for drunkenness for an adequate picture of the growth of intemperance throughout the country.

A brief summary of federal prohibition enforcement statistics for each fiscal year from 1920 to 1928, inclusive, indicates the rapid growth of the illicit liquor industry. The number of pieces of distilling apparatus seized has increased from 95,933 for the first full year of prohibition to 261,611 in 1928. The total amount of liquor seized has increased from 5,805,000 gallons in 1921, the first full year, to 32,474,000 in 1928. Federal arrests of prohibition violators show a similar increase. In 1921, 34,175 persons were arrested for such offenses. With the exception of a slight drop during 1925 and 1926, there has been a steady rise in the number of arrests, culminating in 75,307 in 1928. Convictions in federal courts have increased from 17,962 in 1921 to 58,813 in 1928. This is a more significant figure than the increase in arrests.

To keep pace with the steadily increasing volume of prohibition cases, our federal district courts have been reduced to the status of police courts. In 1928, convictions of prohibition violators accounted for 70.4 percent of all convictions secured in federal courts. The civil cases growing out of prohibition (chiefly padlock injunctions) accounted for 46.3 percent of all civil cases tried in federal courts. Resort

TABLE I

A SUMMARY OF PROHIBITION ENFORCEMENT
(From Annual Reports of Prohibition Unit)

Year Ended June 30th	Arrests	Convictions	Seizures	
			Distilling Apparatus	Liquor—All Kinds
1920	10,548	4,315	15,416	153,735
1921	34,175	17,962	95,933	5,805,297
1922	42,223	22,749	111,155	8,622,230
1923	66,936	34,067	158,132	14,346,632
1924	68,161	37,181	159,176	15,827,189
1925	62,747	39,072	172,537	18,716,259
1926	58,391	41,154	161,979	28,741,810
1927	64,986	36,546	208,073	29,170,831
1928	75,307	58,813	261,611	32,474,234

to "bargain day" methods in prohibition cases is now an established procedure, for 90 percent of the convictions in these cases are secured through pleas of guilty. In the greater number of these cases the defendant agrees to plead guilty, on the assurance of the court and prosecutor that a fine and not a jail sentence will be imposed.

The increase in the conviction of prohibition violators and in the amount of contraband liquor seized may indicate either more effective enforcement or that the illicit traffic is outstripping all efforts at control. The fact that all available indexes of intemperance—deaths from alcoholism, arrests for drunkenness, alcoholic insanity, and acute alcoholism—show a corresponding increase, indicates that so far, at least, the enforcement agents are simply taking larger samplings each year, without curtailing or preventing the spread of the illicit traffic.

Deaths from Alcoholism

The steady increase in the supply of contraband liquor is reflected in the rise in the deaths from alcoholism since 1920. In the two years before national prohibition, the United States Census Bureau reported a sharp drop in the alcoholic death rate. This decrease continued through the first year of prohibition, but beginning with 1921 there has been a steady rise. In 1920, the rate for the 1910 registration states was 1.2 per 100,000; in 1927 it was 5.0, an increase of 318 percent. The 1927 rate is 79.3 percent of the 1916 rate, the peak year before prohibition.

Arrests for Drunkenness

There is no official tabulation of arrests for drunkenness for the entire country. The Moderation League has compiled the annual figures of arrests for drunkenness for over five hundred communities. The figures of the Moderation League have been criticized because no allowance has been made for the increase of population which has taken place from the period 1914 to 1927. From the records of the Moderation League we have compiled arrests for drunkenness for 385 towns and cities for all years from 1914 to 1927 and have calculated the rate per 10,000 population. The annual population was estimated on the basis of the arithmetical increase from 1910 to 1920. We believe these figures are representative of the entire country, for they include about 26 percent of the entire population of the country and 51.4 percent of the urban population of the country. Chicago was omitted from our tabulation because arrests for drunkenness in that city are included with arrests for disorderly conduct.

Table II on the following page gives the rate of arrests for these 385 places. We have also subdivided these cities and towns into four groups, according to size, and have given the rate of arrests for each group of cities. For the entire group the rate of arrests for drunkenness shows a decrease from the peak year of 1916 from 193 per 10,000 to 71 per 10,000 in 1920. Since 1920, there has been a sharp rise up to 1923. From 1923 to 1925, the figures are almost stationary, followed by a slight rise in 1926 and 1927. The 1927 arrests show an increase of 106 percent over 1920, the lowest year. The rate of increase has not been uniform. In towns under 20,000 and in towns between 50,000 and 250,000, there has been a slight decrease since 1923, while in towns from 20,000 to 50,000 and in cities over 250,000 there has been no decrease since 1923, although the increase has been much less marked.

TABLE II

ARRESTS FOR DRUNKENNESS AND RATES PER 10,000 POPULATION IN 385 PLACES, 1914 TO 1927

Year	All Places		Under 20,000 Population		20,000–50,000 Population		50,000–250,000 Population		Over 250,000 Population	
	Arrests	Rate	Arrests	Rate	Arrests	Rate	Arrests	Rate	Arrests	Rate
1914	465,753	187	45,871	218	50,433	197	117,567	198	251,882	176
1915	465,730	183	44,665	209	53,141	203	119,979	197	247,945	170
1916	502,477	193	43,132	199	58,342	217	140,156	225	260,847	175
1917	482,053	182	41,738	189	53,266	193	127,232	199	259,817	171
1918	371,142	137	31,514	141	38,084	135	96,030	147	205,514	133
1919	281,005	102	30,433	134	27,338	95	71,071	106	152,163	96
1920	201,229	71	16,051	69	20,867	71	61,751	90	102,560	64
1921	266,576	93	20,779	89	27,226	91	82,829	119	135,742	83
1922	358,599	122	26,059	110	34,599	113	107,393	150	190,548	114
1923	421,631	141	31,449	130	41,341	132	126,051	173	222,790	131
1924	426,309	140	31,616	129	42,961	134	122,007	164	229,725	133
1925	436,743	141	31,316	126	46,035	141	121,410	160	237,982	135
1926	452,186	144	31,204	124	46,995	141	124,837	161	249,150	139
1927	466,806	146	32,051	126	47,425	140	129,449	164	257,881	142

We believe that deaths from alcoholism are a clearer index of intemperance than arrests for drunkenness, for the latter are affected to a very great extent by changes in police policy. If there were a uniform and consistent police policy with regard to arrests for drunkenness, the two curves should approximate each other. We find, however, that since 1923 the arrests for drunkenness curve has tended to flatten out, while

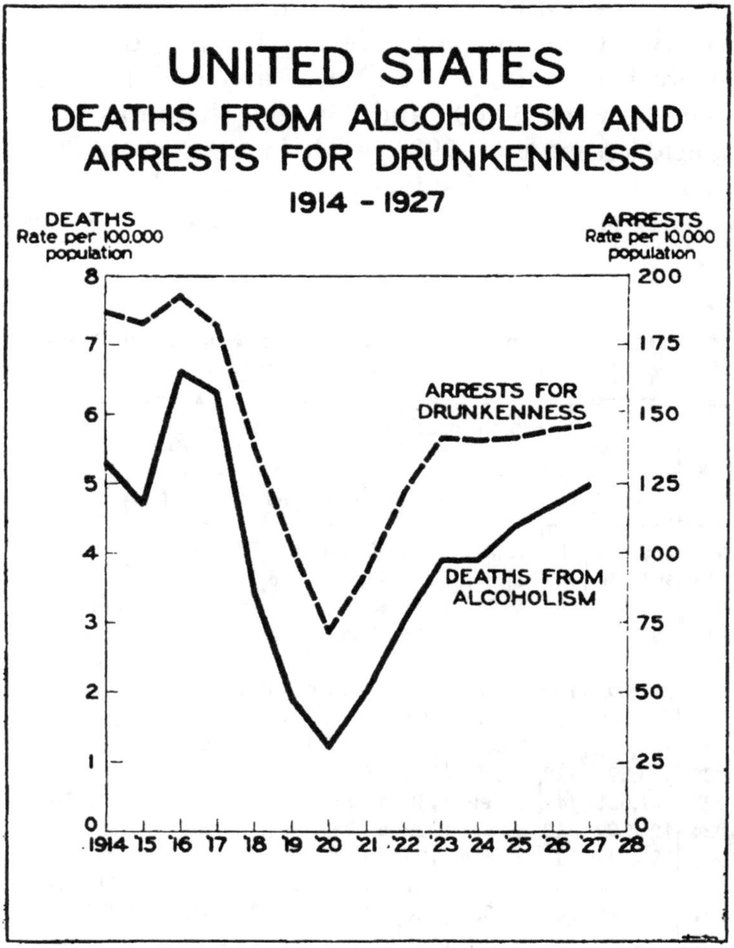

the curve for alcoholic deaths has continued to rise. This is shown graphically in the accompanying chart, which gives the death rate from alcoholism for the registration states of 1910 and the rate of arrests for drunkenness for 385 places.

Arrests for drunkenness vary greatly among different cities because of variations in police policy. For example, in New York City, in 1927, only 11,997 persons were arrested for drunkenness, while in Boston, which is roughly one-eighth the size of New York, 38,794 were arrested for drunkenness. In Pittsburgh, approximately one-tenth the size of New York, there were 31,759 arrests. In Detroit, which is about one-fourth the size of New York, 28,804 were arrested. The differences can hardly be explained on the basis of greater sobriety in New York City.

Professor Fisher has devised what he calls a "statistical corrector" for increased severity in arrests for drunkenness. In his book, *Prohibition Still At Its Worst*, Professor Fisher states that he sent a questionnaire to police departments of 1,000 towns and cities. He received 320 replies. He does not say what questions were asked, nor does he tabulate the replies. He merely says: "A statistical picture of strictness was very conservatively estimated from the police reports, indicating that before prohibition 55 percent of the cases of drunkenness were arrested, and 70 percent for the years 1924, 1925, and 1926." He then applies this "corrector" to the arrest figures which he had collected for these 320 cities. He is able to show "a decline of 122 points in the peak of the period of national prohibition."

While Professor Fisher does not disclose the items covered in his questionnaire, we must infer that he asked the police chiefs whether or not, in their opinion, they now arrest a larger proportion of all drunks than they did before prohibition. This form of questionnaire violates an elementary principle of sound statistical procedure. For statistical purposes, the questions should call for objective answers, and not subjective impressions. Questions should be capable of being answered by a straight "yes" or "no," and leave no room for personal bias or mere opinion. The form of the questionnaire and the opinions on which the "corrector" is based are faulty in both of these respects. Moreover, the short official tenure of most police chiefs precludes them from forming any valid opinion regarding arrest policies before prohibition. We are told by a member of the research staff of the International Association of Chiefs of Police that the average tenure of office of police chiefs is eighteen months. Only in rare instances has the same chief been in office for both the prohibition and pre-prohibition period.

We believe that no satisfactory statistical correction can be made, either for increased or decreased severity, since prohibition. The evidence which we have secured, however, we believe points to the opposite tendency from that claimed by Professor Fisher. A reference to the chart and table which give the deaths from alcoholism and the arrests for drunkenness indicates that the severity was greater during the early period of prohibition, and that since 1923, with alcoholic death rates increasing, arrests for drunkenness have tended to decrease. A rather careful study of the police records of three important cities—Detroit, New York, and Pittsburgh—indicates some of the factors which determine police policy with regard to drunkenness.

Detroit: In 1914, the Detroit Police Department

adopted the "Golden Rule" policy. Drunken persons were sent home, when possible; if incapable of taking care of themselves, they were locked up for safekeeping and set loose after they had sobered up; only those who had been disorderly, or who were charged with some offense besides intoxication, were taken to court. With the advent of state prohibition, the "Golden Rule" was rescinded, on the theory that the country would be bone-dry and that intoxication cases would not hamper the department. After seven years of this policy, the increasing number of arrests for drunkenness made it necessary to go back to the "Golden Rule," which was revived by Commissioner Rutledge, under general order, December 14, 1926.

In his official memorandum Commissioner Rutledge said:

> Commencing this date we will resume our former "Golden Rule" policy in dealing with drunks, in order that several hundred policemen may be saved from spending all day in court prosecuting such misdemeanants, and be afforded the opportunity to demonstrate their ability to prevent serious crimes in Detroit by either sending criminals to jail or to the morgue.
>
> The courts have recently been clogged with unimportant cases which could easily be handled and have been taken care of in the past by precinct officers without consuming the time of the Judges, whose services are required for the disposition of vicious criminals apprehended by our officers. You are instructed to remember and abide by the rule in the Police Manual, Rule 298, and "settle" as many minor cases as possible out of court.

The record of arrests and court cases, presented in the following table and in the accompanying chart, graphically illustrates the increased burden thrown on the courts when the "Golden Rule" policy was suspended, and the marked relief when it was reinstated. Moreover, total arrests in 1927 and 1928 show an increase from 16,516 in 1926 to 23,003 in 1927 and to

28,804 in 1928. Detroit can afford to be more strict in arrests for drunkenness when only 22.4 percent of the cases are taken to court.

TABLE III

ARRESTS FOR DRUNKENNESS—"GOLDEN RULE DRUNKS" AND COURT CASES—DETROIT

Year	Total Arrests for Drunkenness	Drunks—"Golden Rule"	Court Drunks *
1913	10,558		10,558
1914	9,404	7,603	1,801
1915	11,444	9,335	2,109
1916	15,725	12,014	3,711
1917	19,327	15,723	3,604
1918	13,469	11,518 **	1,951
1919	7,030		7,030
1920	6,599		6,599
1921	7,220		7,220
1922	10,098		10,098
1923	12,977		12,977
1924	13,717		13,717
1925	16,298		16,298
1926	16,516	710 †	15,806
1927	23,003	17,721	5,282
1928	28,804	21,785	7,019

* Includes motor drunks since 1918.

** Persons released when sober without the necessity of going to court. The practice was discontinued May 1, 1918, and reinstated December 14, 1926.

† For 17 days only.

New York City: New York never officially adopted the "Golden Rule" policy. The police records for a period of twenty-eight years indicate a clear tendency to reduce arrests to a level which will involve a minimum burden on the courts and penal institutions and a minimum time of the police in taking such cases to court.

In Table IV (page 122) it appears that for the period

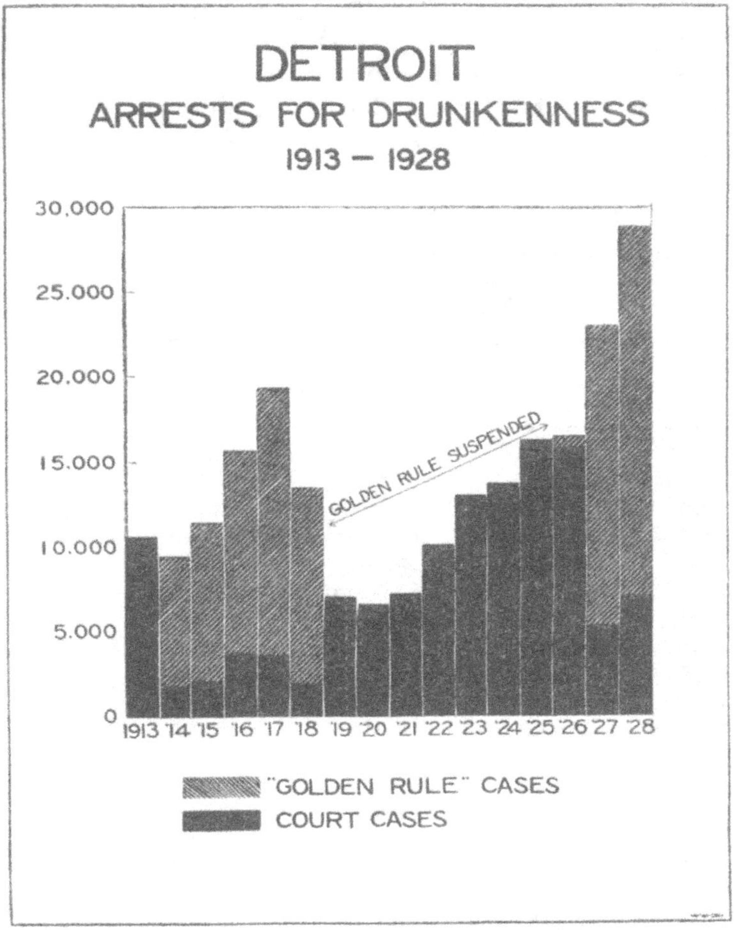

from 1900 to 1907 there was a wide disparity between the number of persons arrested and the number actually arraigned in the magistrates' courts; and of those actually arraigned before the courts, a relatively small proportion was convicted by the magistrates. It was evident, therefore, that during this early period, many drunks were arrested, but were released without the formality of an appearance before the magistrate.

From this point on, however, there is a marked reduction in the number of arrests and also in the number of arraignments, while convictions increased. Since 1916, the number arrested for drunkenness parallels very closely the total number convicted. There was

TABLE IV

ARRESTS, ARRAIGNMENTS, CONVICTIONS, AND DEATHS FROM ALCOHOLISM IN NEW YORK CITY
(From World Almanac, 1929)

Year	Arrests	Arraignments	Convictions	Deaths from Alcoholism
1900	42,012	17,642	10,781	402
1901	42,836	18,801	10,937	329
1902	46,572	19,927	12,170	345
1903	53,396	No Record	No Record	630
1904	50,933	27,343	13,457	665
1905	52,316	No Record	No Record	596
1906	47,749	23,748	9,376	636
1907	44,787	23,016	9,364	719
1908	No Record	20,292	8,403	409
1909	36,668	33,367	22,372	533
1910	35,258	27,310	19,106	621
1911	26,767	20,599	14,796	636
1912	25,605	19,123	14,232	570
1913	26,207	20,848	16,600	656
1914	23,122	19,936	16,728	660
1915	22,635	19,587	17,773	562
1916	19,453	17,078	15,684	687
1917	16,311	14,182	12,913	560
1918	8,795	7,284	6,878	252
1919	7,028	5,657	5,372	176
1920	7,804	6,340	6,061	98
1921	8,169	6,726	6,485	119
1922	11,402	9,309	8,887	274
1923	14,018	11,192	10,605	429
1924	13,980	11,256	10,744	513
1925	12,017	9,391	8,970	682
1926	12,330	9,533	9,120	759
1927	11,997	8,692	8,341	813

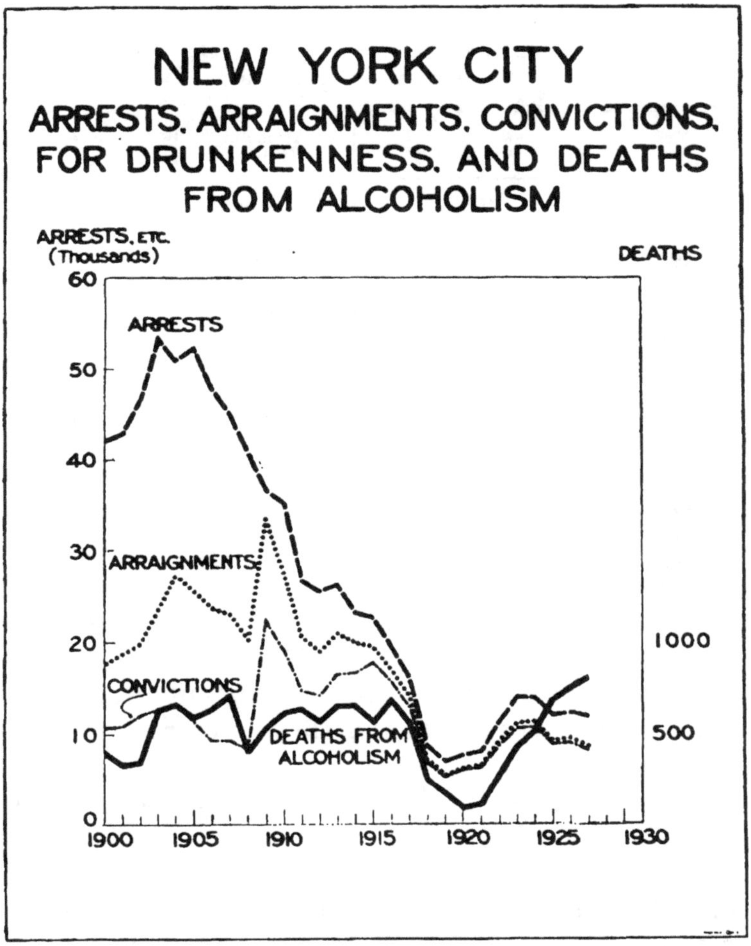

a sharp increase in the number of persons convicted in 1909. Convictions in that year were 22,372, as compared with 8,403 in 1908. This increase in convictions is to be attributed, we believe, to the reorganization of the magistrates' courts, which took place at that time, and for several years resulted in a rather high rate of convictions. The number of convictions gradually trailed off, however, until in

TABLE V

RATIO OF ARRESTS FOR DRUNKENNESS TO DEATHS FROM ALCOHOLISM—PITTSBURGH

Year	Arrests for Drunkenness	Alcoholic Deaths	Ratio
1913	21,970	No Record	
1914	20,567	56	367
1915	21,990	69	319
1916	27,376	85	322
1917	28,935	103	281
1918	23,556	38	620
1919	14,304	21	681
1920	9,577	17	563
1921	10,371	26	399
1922	16,556	31	534
1923	24,651	67	368
1924	25,401	83	306
1925	28,669	113	254
1926	28,427	95	299
1927	34,070	107	318
1928	31,759	102	311

1916 the numbers of arrests, arraignments, and convictions were very close, and they have remained so ever since.

The deaths from alcoholism for this period serve as a rough check on police severity. The trend of deaths before 1920 is much the same as that of convictions, except for the period for 1909 to 1916, when the courts were unusually severe. Since 1920, the death rate from alcoholism has increased 683 percent, while the arrest rate has increased only 45 percent. As a matter of fact, in recent years the number arrested for drunkenness barely exceeds the number of patients committed to city hospitals for acute alcoholism. From 1925 to 1927, 36,344 persons were arrested for drunkenness and 32,297 patients were committed to city hospitals for acute alcoholism. Apparently only

10 percent of those arrested for drunkenness are still "ambulatory," and 90 percent are "bed" cases.

Pittsburgh: If deaths from alcoholism may be used as a check of police severity, the figures we have collected for Pittsburgh (Table V) are of interest. From 1914–1916, 333 persons were arrested for drunkenness for each death from alcoholism. The ratio increased until in 1920–1922, 493 were arrested for each alcoholic death. Since 1923, the ratio has varied around 300, a little below the pre-war level. Apparently the peak of severity was between 1917 and 1922.

Liquor Curves in Foreign Countries

In the evidence so far presented, we have assumed that arrests for drunkenness and deaths from alcoholism, particularly the latter, are an index of liquor consumption. The records of foreign countries which have dependable data bear out this assumption. Not only is liquor consumption declining in other countries, but arrests for drunkenness and deaths from alcoholism show a similar decline.

The charts which give liquor consumption and, where available, arrests for drunkenness and alcoholic deaths in foreign countries, illustrate not only the close correlation between liquor consumption and indexes of intemperance, but they show that in foreign countries, regardless of the particular form of liquor regulation, the tendency both in liquor consumption and in intemperance is consistently downward. Moreover, except in Finland, which has had prohibition since 1919, the post-war level of consumption and intemperance is far below the pre-war level.

Dr. Clark Warburton, who is making a study of the economic effects of prohibition, has worked out the coefficients of correlation of liquor consumption with

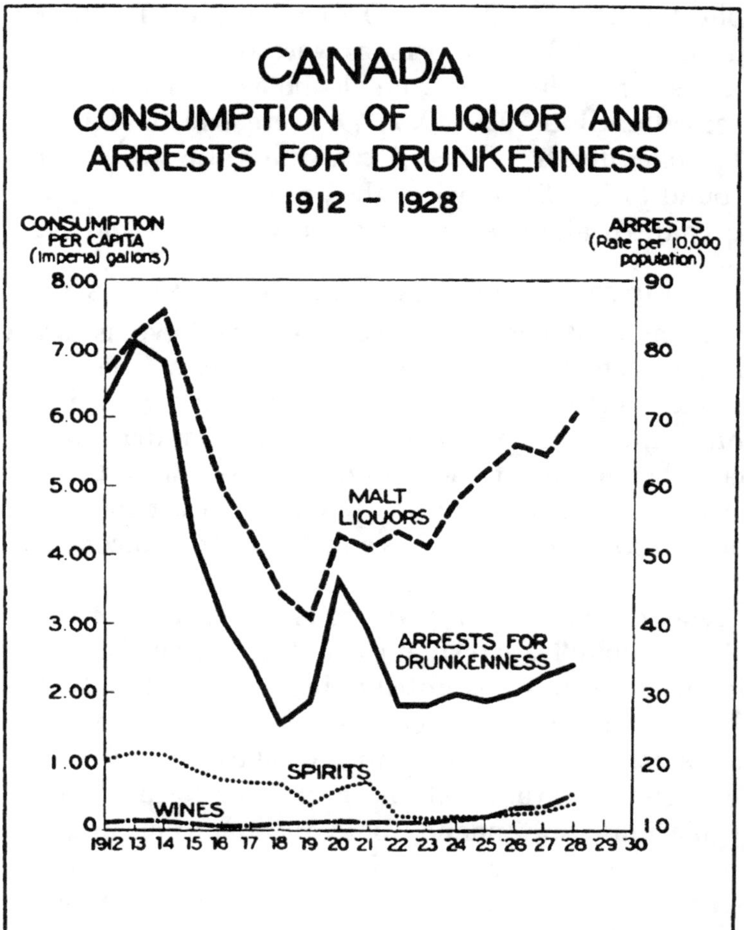
CANADA
CONSUMPTION OF LIQUOR AND
ARRESTS FOR DRUNKENNESS
1912 - 1928
CONSUMPTION
PER CAPITA
(Imperial gallons)
ARRESTS
(Rate per 10,000
population)
8.00
7.00
6.00
5.00
4.00
3.00
2.00
1.00
0
90
80
70
60
50
40
30
20
10
MALT
LIQUORS
ARRESTS FOR
DRUNKENNESS
SPIRITS
WINES
1912 13 14 15 16 17 18 19 '20 21 '22 23 '24 '25 '26 27 '28 '29 '30

DEATHS, CONVICTIONS AND CRIMES OF VIOLENCE DUE TO DRUNKENNESS IN SWEDEN

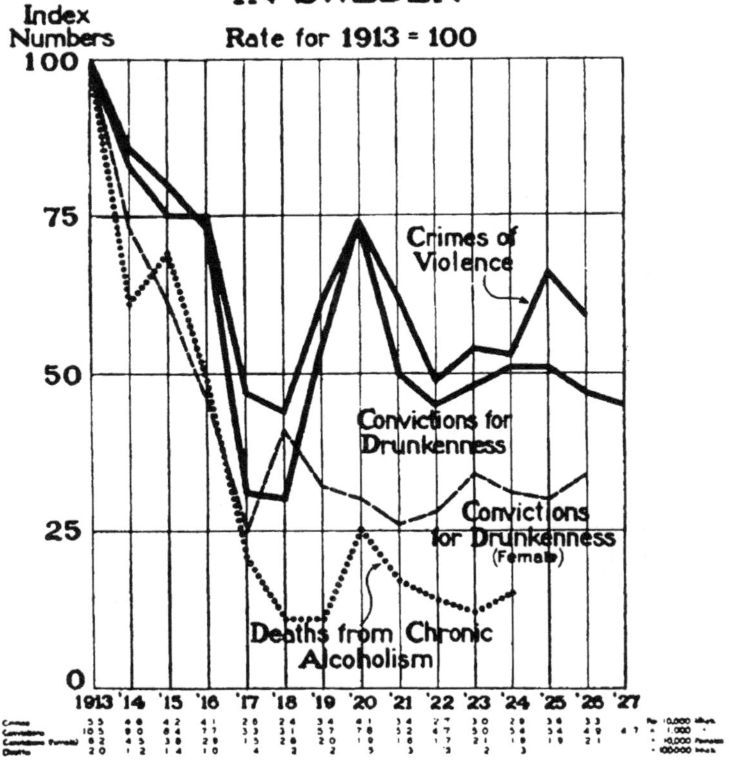

ARRESTS, CONVICTIONS AND CRIMES OF VIOLENCE DUE TO DRUNKENNESS IN STOCKHOLM

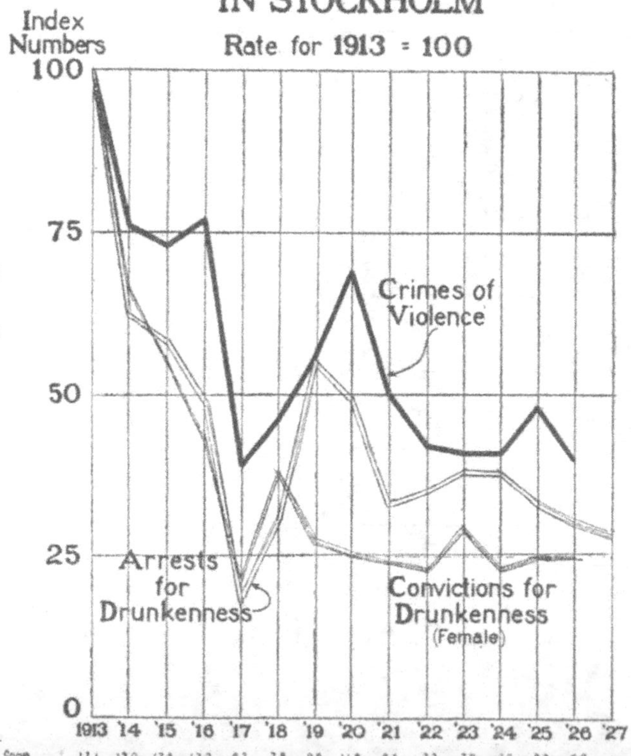

CHRONIC ALCOHOLISM CASES TREATED IN KATARINA HOSPITAL STOCKHOLM

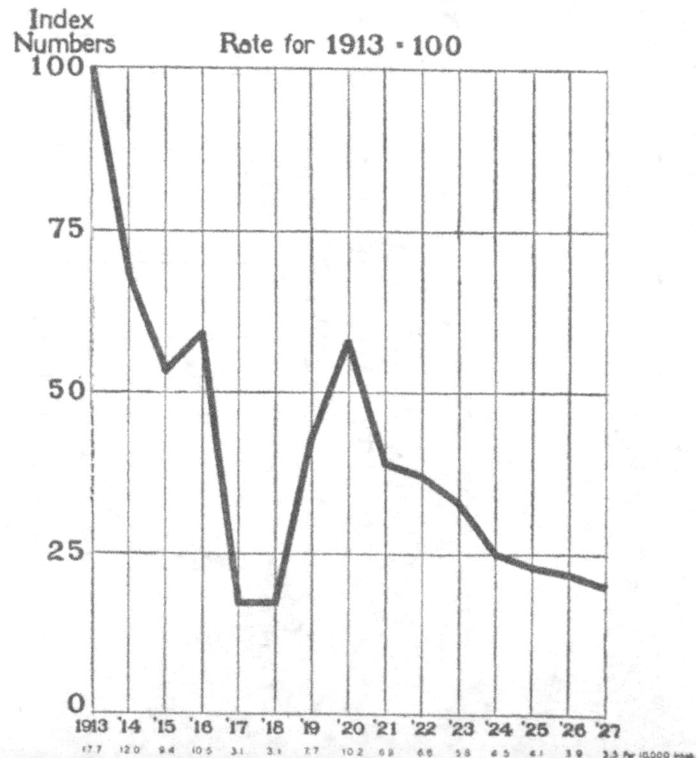

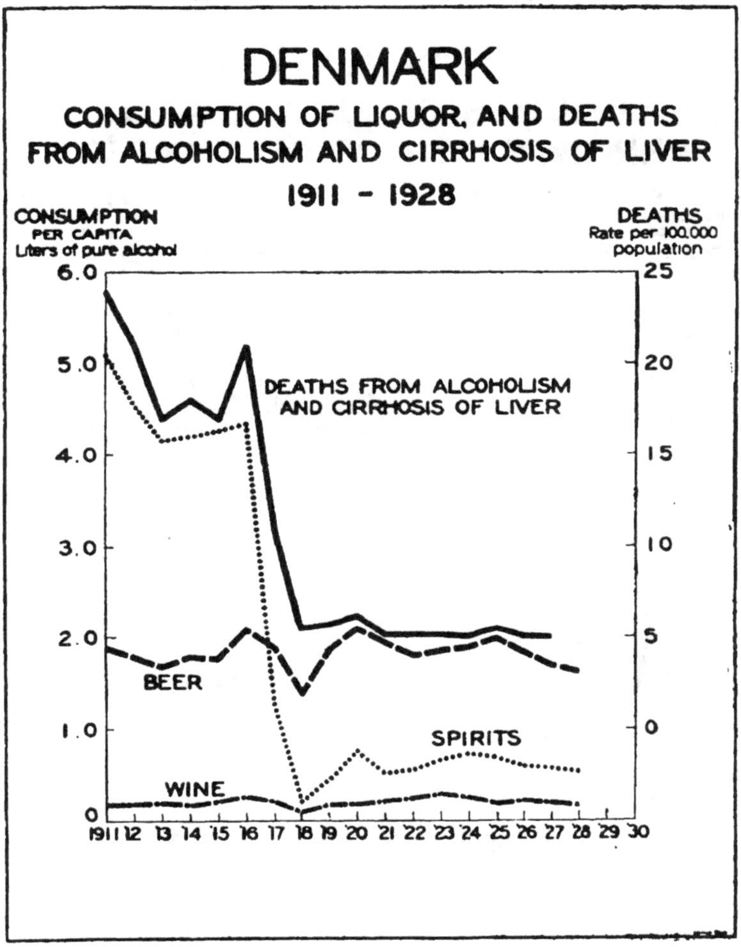
DENMARK
CONSUMPTION OF LIQUOR, AND DEATHS FROM ALCOHOLISM AND CIRRHOSIS OF LIVER
1911 - 1928
CONSUMPTION PER CAPITA Liters of pure alcohol
DEATHS Rate per 100,000 population
DEATHS FROM ALCOHOLISM AND CIRRHOSIS OF LIVER
BEER
SPIRITS
WINE
6.0
5.0
4.0
3.0
2.0
1.0
0
25
20
15
10
5
0
1911 12 13 14 15 16 17 18 19 20 21 22 23 24 25 26 27 28 29 30

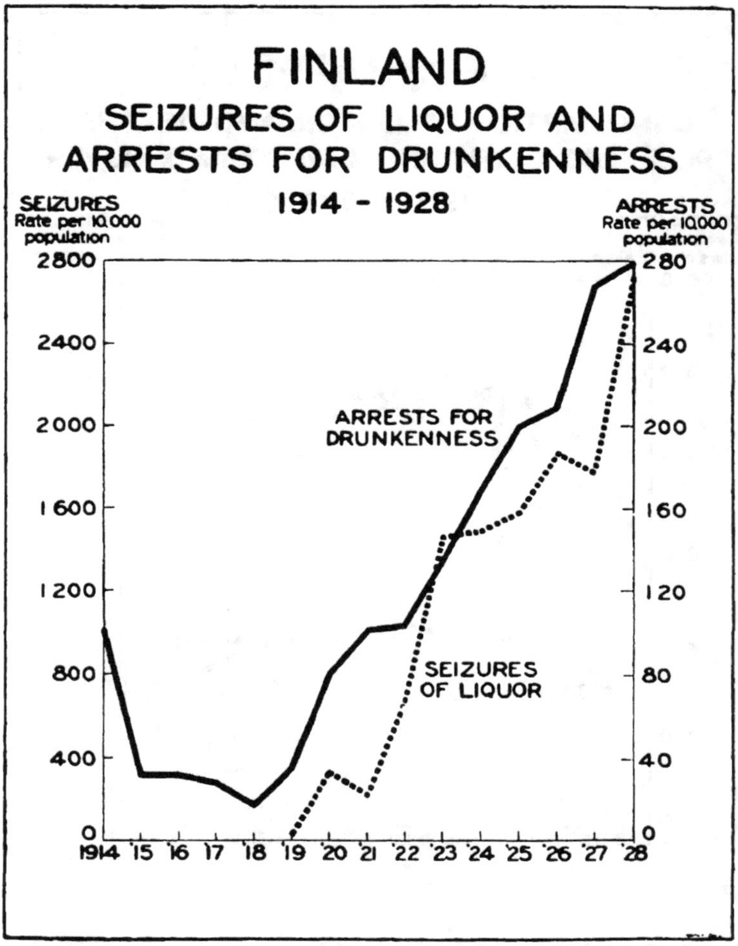
FINLAND
SEIZURES OF LIQUOR AND
ARRESTS FOR DRUNKENNESS
1914 - 1928
SEIZURES
Rate per 10,000
population
ARRESTS
Rate per 10,000
population
2800
2400
2000
1600
1200
800
400
0
280
240
200
160
120
80
40
0
ARRESTS FOR
DRUNKENNESS
SEIZURES
OF LIQUOR
1914 '15 '16 '17 '18 '19 '20 '21 '22 '23 '24 '25 '26 '27 '28

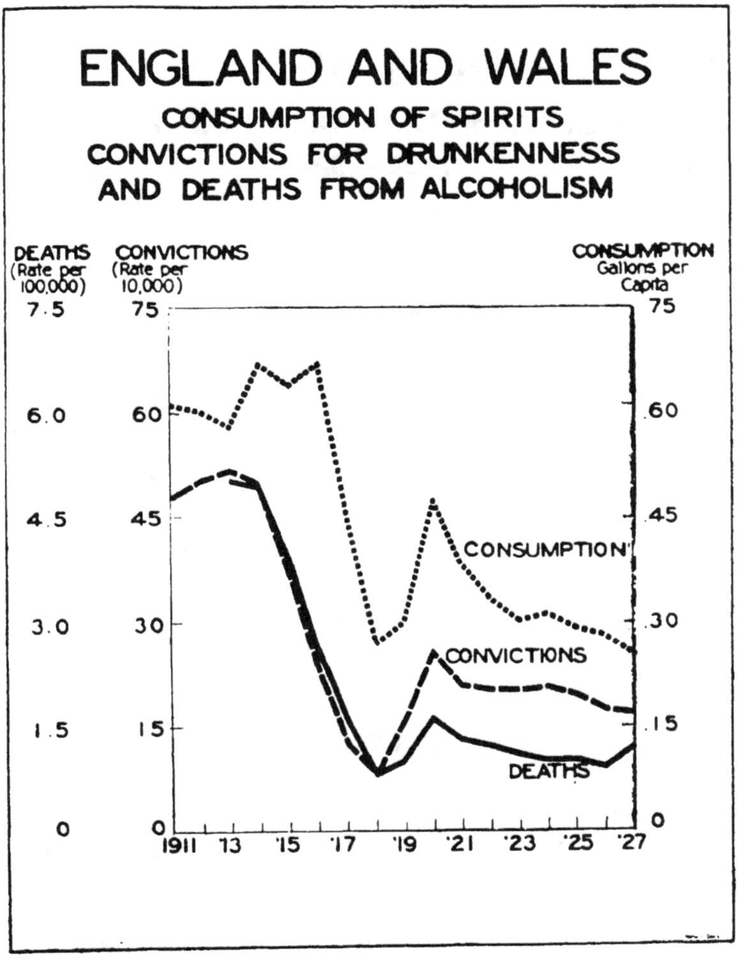
ENGLAND AND WALES
CONSUMPTION OF SPIRITS
CONVICTIONS FOR DRUNKENNESS
AND DEATHS FROM ALCOHOLISM
DEATHS (Rate per 100,000)
CONVICTIONS (Rate per 10,000)
CONSUMPTION Gallons per Capita
7.5
6.0
4.5
3.0
1.5
0
75
60
45
30
15
0
75
.60
.45
.30
.15
0
CONSUMPTION
CONVICTIONS
DEATHS
1911
'13
'15
'17
'19
'21
'23
'25
'27

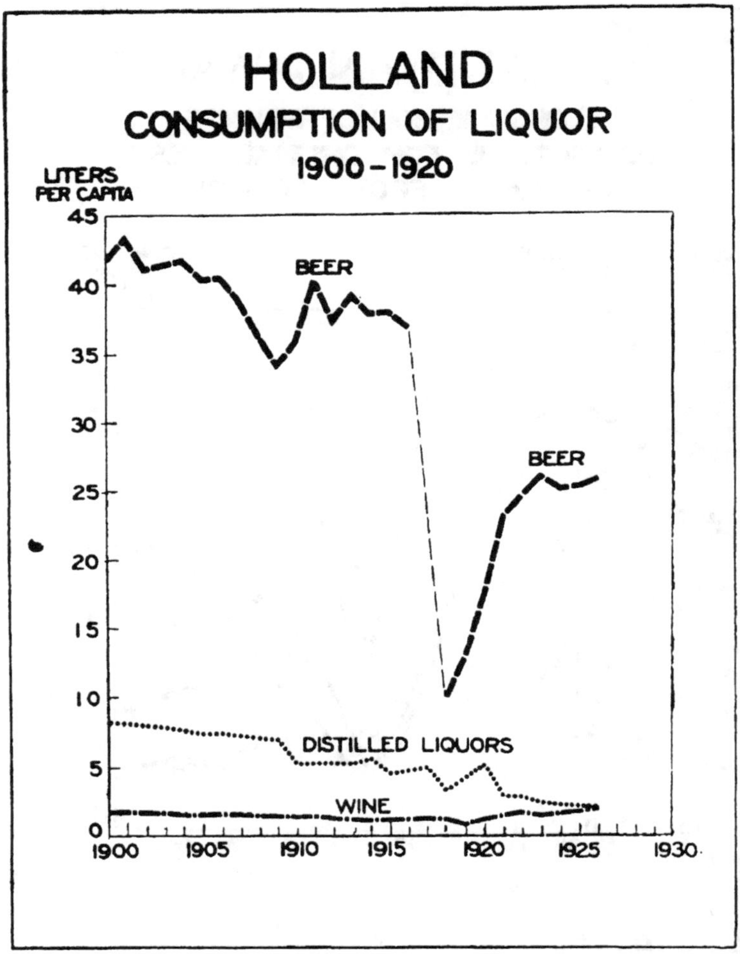
HOLLAND
CONSUMPTION OF LIQUOR
1900-1920
LITERS PER CAPITA
45
40
35
30
25
20
15
10
5
0
BEER
BEER
DISTILLED LIQUORS
WINE
1900
1905
1910
1915
1920
1925
1930

alcoholic death rates in the United States and various European countries. He finds the correlation to be amazingly high. Dr. Warburton has tested various kinds of "curve-fitting." He has correlated consumption of each type of liquor with deaths from alcoholism, with cirrhosis of the liver, and with a combined rate of both, and with and without a half-year lag. So far he has worked out fifty-seven correlation rates in the United States, Great Britain, Denmark, and Sweden. The correlation is lowest in Sweden, .656 ± .111. This is because for a three-year period legal consumption figures are unreliable, and for the entire series alcoholic deaths are unreliable. With this exception, the correlation coefficients range from .834 to .974. Indeed, it is not unlikely that we shall be able to calculate from alcoholic death rates the probable consumption of liquor in those countries, like the United States, where official consumption records are lacking.

Economic Effects of Prohibition

A report on the social effects of prohibition would be incomplete without some reference to the economic aspects of the problem. For a year and a half, we have been making a study of the economic effects of prohibition. We can only give here the main conclusions of our study, and these only in tentative form.

It is believed by many, even by some who are in principle opposed to the Eighteenth Amendment, that prohibition has made a substantial contribution to the economic welfare of the country. It is claimed that most of the money formerly spent on liquor now goes to the purchase of other commodities, socially or economically more productive than drink, or into savings. It is also claimed that the reduction in the consumption of alcohol has increased the productivity of the American worker and so contributed to our national wealth.

Present Consumption of Liquor

No one knows how much liquor is now being consumed in the United States. From official and trade sources we have estimated, within reasonable limits, the materials which go into the manufacture of intoxicating beverages. From the total production of materials, we have deducted the amounts used in legitimate industries. This leaves the amount which probably goes into the manufacture of intoxicating beverages. For wine and beer this process is fairly simple and reliable. In the case of spirits, however, many baffling problems are presented.

Beer: Our estimate of beer production is based on the hop crop. In Table VI are given the annual hop crops for 1926 to 1928, with deductions for net exports and the amounts used by manufacturers of non-intoxicating cereal beverages. This leaves a balance which is used almost exclusively for real beer.

The average annual balance for the three years is sixteen million pounds, and the trend is decidedly upward. The preliminary estimate for the 1929 hop crop given by the United States Department of Agriculture is 33,220,000 pounds.

We believe it can be conservatively estimated that at least 15,000,000 pounds of hops are used annually in the making of real beer. We have put the figure

TABLE VI

HOP CONSUMPTION IN THE UNITED STATES
(Thousands of Pounds)

Year	Last Year's Production	Net Exports	Used by Beverage Manufacturers	Balance (Probably Used to Make Beer)
1926	28,573	12,265	3,426	12,882
1927	31,522	13,565	3,148	14,809
1928	30,658	7,404	3,071	20,183

low because we do not have complete records of carry-over of hops from year to year.

Before prohibition brewers used from .64 to .71 pounds of hops for each barrel of beer, according to the reports of the Bureau of Internal Revenue. Because of more wasteful methods today, probably .75 pounds are used for each barrel of beer. Assuming that at least 15,000,000 pounds of hops are used, we may safely conclude that 20,000,000 barrels, or 640,000,000 gallons of real beer are made annually. For practical purposes we may disregard some four or five million gallons of beer smuggled from Canada and Mexico.

Wine: In our estimate of the amount of grapes used in wine-making, we have deducted from the total crop twenty-five pounds per capita for all other uses. This is a liberal allowance, for in 1917 the per capita figure for table use and unfermented juice was less than nineteen pounds.

The following table gives two authoritative estimates of the per capita consumption of grapes for table use and unfermented beverages before prohibition. They are taken from the United States Census of Agriculture in 1910 and Shear's *Economic Status of the Grape Industry.*

TABLE VII

ESTIMATED PER CAPITA CONSUMPTION OF GRAPES IN THE UNITED STATES FOR OTHER PURPOSES THAN WINE AND BRANDY MAKING

Year	Estimate No. 1 (Pounds)	Estimate No. 2 (Pounds)
1899	8.8	7.8
1909	12.9	9.6
1917	18.4	16.7

We have assumed that the per capita consumption of grapes for table use and unfermented grape juice has increased since 1917 to twenty-five pounds, which is equivalent to sixty percent of the crop. There is, however, no evidence that the per capita rate has increased to that extent.

Shear estimates that 150 gallons of wine can be made from one ton of grapes. The following table gives a summary of our estimate of the annual wine production from 1926 to 1928.

TABLE VIII

ESTIMATE OF WINE PRODUCTION IN THE UNITED STATES, 1926–1928

Year	Production (Thousand Tons)	Table Use (Thousand Tons)	Balance (Thousand Tons)	Wine Produced (Million Gallons)
1926	2,423	1,464	959	144
1927	2,463	1,483	980	147
1928	2,483	1,500	983	147

To the estimate of home-made wine, we have added roughly 3,000,000 gallons of wine from other sources, including an estimate of 500,000 gallons smuggled into the country and 2,250,000 gallons legally sold for sacramental purposes.

Spirits: The problem of estimating the amount of

illicit spirits produced is exceedingly difficult because it is derived from a wide variety of materials. The chief sources, however, are corn sugar, grain mash, cane and beet sugar, molasses, malt syrup, and illegally diverted denatured alcohol.

Corn Sugar: During recent years corn sugar, or solid glucose, has become one of the chief sources of material from which spirits are made. Distillation of corn sugar has become popular because it is a cleanly process, emits no odor, and leaves no ash. Stills using corn sugar are, therefore, less likely to be detected than those using grain mash.

Corn sugar production has increased from 152,000,000 pounds in 1921 to 905,000,000 in 1927. The per capita production was declining prior to 1921. The following are the official figures of corn sugar production, taken from the United States Statistical Abstract:

TABLE IX

CORN SUGAR PRODUCTION

Year	Production (Million Pounds)	Pounds Per Capita
1910	174.4	1.78
1919	157.0	1.50
1921	152.0	1.41
1923	527.9	4.73
1925	880.4	7.63
1927	904.8	7.63
1928	968.6 *	8.07 *

* Estimated (from trade sources).

We have attempted to allow for all legitimate uses of corn sugar before arriving at the balance which is used in making illicit spirits. We have allowed for an increase in legitimate pre-prohibition uses, from 1.5

pounds to 1.8 pounds per capita. To this we have added 70,000,000 pounds used annually in the rayon industry. This industry has had its greatest growth since 1919. This leaves from 640,000.000 to 685,000,-000 pounds to be accounted for.

The estimate of the amount of corn sugar used in the rayon industry was made by an executive officer of a company which makes about half of the rayon produced in this country. We could find no other new uses which would account for the tremendous increase in this product. Corn sugar, or solid glucose, is not to be confused with corn syrup, which can be pretty fully accounted for in legitimate industry; nor with corn starch, an entirely different product, which is used in making laundry starch and other legitimate products.

Chemists estimate that about 12.5 gallons of spirits can be made from 100 pounds of corn sugar. This would produce from 80,000,000 to 85,000,000 gallons of spirits, if the quantities of corn sugar indicated above are available.

Industrial Alcohol: In 1926, Commissioner Doran estimated that 5,000,000 gallons of completely denatured alcohol and 8,000,000 gallons of specially denatured alcohol were diverted in 1925 for illicit beverage alcohol. We have reduced Commissioner Doran's total estimate to a range of from 2,500,000 gallons to 7,500,000 gallons of denatured alcohol as diverted in 1927. This would produce from five to fifteen million gallons of proof spirits.

Molasses: Unless the per capita legitimate uses of molasses have declined, this is not an important source of beverage alcohol. A downward trend of .2 or .3 gallons per capita would yield thirty-five million gallons available for illicit distillation. This would produce twenty-five million gallons of proof spirits.

The following table gives the production of molasses and the amount used for the legitimate production of alcohol (chiefly industrial alcohol):

TABLE X

MOLASSES

Year	Production	Used in Legitimate Alcohol Production	Balance	Per Capita
1914	133.3	64.7	68.6	.7
1919	196.5	123.5	73.0	.7
1921	139.3	119.0	20.3	.2
1923	262.3	148.7	113.6	1.0
1925	332.8	203.3	129.5	1.1
1927	307.9	211.5	96.4	.8

Cane and Beet Sugar: The extent to which cane and beet sugar are used for illicit liquor depends on the estimate of normal and legitimate use. This was arrived at by two methods. In the first method, a trend line has been fitted to the last ten pre-war years. In the second, the estimate of normal consumption for any year is based on the (*a*) ratio of population in any given year to that of 1913, and (*b*) ratio of wholesale sugar prices to a three-year moving average of the Bureau of Labor Index of Wholesale Prices. According to the first estimate, legitimate use accounts for all the sugar produced. According to the second method, 300,000 tons of sugar, producing 75,000,000 gallons of proof spirits, are available. The latter is a maximum estimate. The truth probably lies between the two extremes.

Malt Syrup: Fox estimates that three gallons of malt syrup are used for every one hundred pounds of corn sugar in production of the Scotch type of whiskey in Detroit. If this holds for the entire country, we should add three percent to our estimates of the quantities of corn and cane sugar used in distilling. This would call for about twenty to thirty million gallons of malt syrup. From this, fifteen million gallons of proof spirits could be made. This is used only in our maximum estimate.

Corn and Other Grains: It is difficult to estimate the amount of spirits which is distilled directly from corn and other grains. Enforcement statistics indicate that the quantity is not decreasing. In 1923, roughly 8,000,000 gallons of mash were seized; by 1926 it had risen to 12,000,000, and by 1928, 26,594,533 gallons were seized. This mash is mostly grain mash, for corn and cane sugar are more difficult to discover and are less bulky. It is reasonable to suppose that the quantity of spirits distilled and consumed from grain mash is at least equal to the amount of mash seized. Our minimum estimate, therefore, is 20,000,000 gallons from this source. Our maximum estimate is that spirits distilled from grain mash are equivalent to those distilled from corn sugar.

If we combine the estimates of production from these sources, we have the resulting minimum and maximum estimates of illicit spirits made and consumed in 1927, as given in Table XI.

The range between these extremes is obviously too great. There is no scientific way of arriving at a mean between them. To be conservative, our final estimate is that, from all sources, at least 180,000,000 gallons of illicit spirits were produced and consumed in 1927. There are many indications that since 1927 consumption has increased.

We have estimated the cost to the consumer of

TABLE XI

PROBABLE CONSUMPTION OF BEVERAGE SPIRITS IN THE UNITED STATES IN 1927
(Gallons, Proof Strength)

	Minimum	Maximum
Medicinal use, and obtained from smuggling	5,000,000	15,000,000
Industrial alcohol diverted	5,000,000	15,000,000
Produced from corn sugar	80,000,000	85,000,000
Produced from molasses	Negligible	25,000,000
Produced from cane and beet sugar	"	75,000,000
Produced from malt syrup	"	15,000,000
Produced from corn and other grains directly	20,000,000	85,000,000
	110,000,000	315,000,000
Per Capita	.93 gal.	2.66 gal.

various types of liquor. Bootleg prices are not representative of the entire country. Home-made beer can be made for less than fifty cents a gallon, and wine can be made from grapes for from eighty cents to one dollar a gallon. Spirits can be made for less than one dollar a gallon. To these minimum estimates we have added an allowance for liquor bought from bootleggers and in speakeasies.

Table XII gives our best tentative estimate of the probable production of illicit liquor, with estimated retail prices.

NET SAVING IN THE DRINK BILL

Before prohibition *The American Grocer*, a trade paper, made reliable estimates of the annual drink bill. Based on their estimates, in 1917, the last typical year, the total drink bill for the country was $1,817,000,000. We have deducted federal and state

TABLE XII

ESTIMATED EXPENDITURE ON ALCOHOLIC BEVERAGES, 1927

	Quantity (gallons)	Per Capita	Minimum		Maximum	
			Price	Expenditure	Price	Expenditure
Beer.......	640,000,000	5.38	$ .50	$ 320,000,000	$ 1.50	$ 960,000,000
Wine.......	150,000,000	1.26	1.00	150,000,000	2.00	300,000,000
Spirits......	180,000,000	1.52	5.00	900,000,000	15.00	2,700,000,000
				$1,370,000,000		$3,960,000,000

revenues, leaving a net drink bill of $1,474,000,000. Allowing for the increase in population, we estimate that if there were no prohibition, the probable drink bill today would be $2,500,000,000.

We believe our minimum estimate of present expenditures is too low, but even this estimate indicates a saving of only $1,130,000,000. Our more liberal estimate would indicate a net loss of over $1,460,000,000. We believe that the truth lies somewhere between these two extremes and that we are spending for liquor today approximately what we should have been spending if there were no prohibition.

The Workers' Share of the Drink Bill

There is no reliable estimate of the proportion of the national drink bill, either before or since prohibition, which may be allocated to the wage-earning population. The United States Department of Labor budget estimates in 1891 and 1918 are obviously understatements of liquor expenditures by workmen's families, for they account for only one-fifth of the average per family expenditure. It is not unlikely that there has been some net gain in temperance among wage-earners through the abolition of the saloon. It cannot be assumed, on the other hand, that the disappearance of the saloon has entirely eliminated excessive drinking among workers.

Probably 1,000,000 persons are arrested every year for drunkenness. Does anyone believe that all of those arrested were millionaire clubmen or captains of industry? The 11,000 acute alcoholic patients treated in the city hospitals of New York largely come from the wage-earning population of the city. The 18,000,000 industrial policy holders of the Metropolitan Life Insurance Company represent a fair cross-section of the wage-earning population. The alcoholic death rate for this group parallels that for the entire country, except for a slight slowing up in the rate since 1926.

Increased Productivity

Extravagant claims have been made as to the effect which prohibition has had on increased productivity of labor. It is extraordinary that the two-volume report of the Committee on Recent Economic Changes, under the chairmanship of Mr. Hoover, which covers practically all of the prohibition period, makes no mention of prohibition as a factor in increased productivity. In that study, Leo Wolman gives an index of output per worker for all years from 1899 to 1927. In the following table we have compared this index for the years 1899 to 1919 with the known liquor consumption, and find little or no correlation. Liquor consumption since 1920 cannot be accurately stated, but intemperance indexes show a steady increase since 1921.

Since 1920, there has been a marked improvement in personnel management, improved machinery, shorter hours of work, and better standards of living. These factors, apart from prohibition, are sufficient to account for the increase in productivity which has taken place, although it is impossible to analyze the correlation mathematically.

Effect of Prohibition on Savings

There has been no phenomenal increase in savings since 1920. Particularly is this true if we exclude time deposits, for these represent in large measure the funds of large corporations. They are not savings in the ordinary sense, although they are usually included in reports of saving deposits. The reduction in the

TABLE XIII

INDEXES OF PRODUCTIVITY AND OF ALCOHOL CONSUMPTION

Year	Output per Person in Manufacture—1911–14 = 100	Index of per Capita Consumption of Alcohol—1911–14 = 100 (From Internal Revenue Bureau)	Annual Change	
			In Index of Output	In Index of Alcohol Consumption
1899	90.3	75.9		
1900	86.9	81.4	– 3.4	5.5
1901	92.0	82.0	5.1	.6
1902	93.4	88.4	1.4	6.4
1903	91.0	90.8	– 2.4	2.4
1904	93.9	92.0	2.9	1.2
1905	102.3	90.5	8.4	– 1.5
1906	103.0	97.1	.7	6.6
1907	98.0	104.1	– 5.0	7.0
1908	91.7	96.8	– 6.3	– 7.3
1909	99.0	92.6	7.3	– 4.2
1910	98.2	97.4	– .8	4.8
1911	93.4	101.1	– 4.8	3.7
1912	103.5	98.0	10.1	– 3.1
1913	105.0	101.7	1.5	3.7
1914	98.0	99.2	– 7.0	– 2.5
1915	106.1	85.9	8.1	–13.3
1916	107.7	89.3	1.6	3.4
1917	99.2	97.1	– 8.5	7.8
1918	94.6	66.6	– 4.6	–30.5
1919	94.4	47.8	– .2	–18.8
1920	97.5	15.1 *	3.1	–32.7
1921	96.9		– .6	
1922	116.1		19.2	
1923	119.7		3.6	
1924	120.1		.4	
1925	131.3		11.2	
1926	134.3		3.0	
1927	135.0		.7	

* Incomplete, legal consumption only.

reserve required for time deposits by the Federal Reserve Board has encouraged banks to solicit time deposits of corporations and large investors, and has

TABLE XIV

TOTAL SAVINGS DEPOSITS IN THE UNITED STATES

Year	Total Savings Deposits June 30 (Millions)	Savings Deposits Per Capita	Annual Percent Increase in Per Capita Savings	Estimated National Income of Preceding Year (Millions)	Increase in Savings as Percent of National Income	Total Individual Deposits Per Capita	Savings Deposits as Percent of all Deposits
1910	$ 6,835	$ 74.08		$30,200		$159	47
1911	7,963	85.00	14.7	32,900	3.4	167	51
1912	8,404	88.38	4.0	32,700	1.3	176	50
1913	8,548	88.57	0.2	34,600	0.4	187	47
1914	8,712	88.96	0.4	35,700	0.5	192	46
1915	8,807	88.66	−0.3	35,600	0.3	187	47
1916	9,459	93.88	5.9	37,200	1.8	218	43
1917	10,876	106.44	13.4	43,300	3.2	223	48
1918	11,535	111.35	4.6	51,300	1.3	236	47
1919	13,040	124.19	11.5	60,400	2.5	271	46
1920	15,314	143.90	15.9	65,900	3.5	303	47
1921	16,501	153.02	6.3	74,000	1.6	317	48
1922	17,579	160.91	5.2	63,400	1.7	322	48
1923	19,727	176.61	9.8	65,900	3.3	366	48
1924	21,189	186.31	5.5	74,300	2.0	366	51
1925	23,134	200.51	7.6	77,100	2.5	399	51
1926	24,696	210.83	5.1	81,900	1.9	405	52
1927	26,091	219.94	4.3	85,700	1.6	414	53
1928	28,413	236.75	7.6	87,500	2.7	427	56
1929	28,208	232.35	−1.9	89,000	−0.2		

led the latter to transfer demand deposits to time deposits. The tendency has been most marked since 1919.

The average annual percentage increase in all savings deposits from 1910 to 1919 was 7.4. From 1920 to 1929 it was 7.1, a slight decline during the prohibition period. A similar decrease is noted in the average annual per capita savings deposits. For the nine years before prohibition they were 6.0. Since prohibition they were only 5.5. In proportion to national income there is a slight increase in savings deposits. Before prohibition, savings deposits were 1.6 percent of annual national income. Since 1920, they represented 1.9 percent.

Deposits in savings banks which represent largely the savings of wage-earners show a slight per capita increase, but show no increase in proportion to national income. For the ten-year period from 1909 to 1919 the average annual increase in per capita deposits was 3.3 percent. For the first eight years of prohibition it was 4.2 percent. The percentage of national income represented by savings banks deposits was exactly the same for both periods, 0.6 percent. This is a more significant figure.

The year 1920 does show a marked increase over 1919 in both series. This may be attributed in part to the boom conditions of that year, in part to the fact that liberty bonds were sold by small investors and part of such funds found their way back to savings banks, and perhaps in part to the fact that liquor consumption was at its lowest ebb in 1920. In any case, the 1920 increase has not been maintained throughout the prohibition period.

TABLE XV

Savings Deposits in Savings Banks and in Federal Reserve Member Banks

Year	Deposits in Savings Banks June 30 (Millions)	Deposits Per Capita in Savings Banks	Annual Percent Increase in Per Capita Deposits	Estimated National Income of Preceding Year (Millions)	Increase of Deposits in Savings Banks as Percent of National Income	Time Deposits All Federal Reserve Member Banks (Millions)	Percent Time Deposits are of All Deposits Federal Reserve Member Banks
1910	$ 4,070	$44.12	7.7	$30,200	1.2		
1911	4,213	44.97	1.9	32,900	0.4		
1912	4,452	46.81	4.1	32,700	0.7		
1913	4,725	48.96	4.6	34,600	0.8		
1914	4,934	50.38	2.9	35,700	0.6		
1915	4,997	50.30	0.1	35,600	0.2	$ 1,352	16.6
1916	5,090	50.52	0.3	37,200	0.3	1,775	17.7
1917	5,419	53.03	5.0	43,300	0.8	2,304	19.2
1918	5,472	52.82	−0.4	51,300	0.1	3,395	21.7
1919	5,903	56.22	6.4	60,400	0.7	4,344	22.8
1920	6,538	61.44	9.3	65,900	1.0	5,911	26.4
1921	6,970 *	64.58	5.1	74,000	0.6	6,367	30.7
1922	7,181	65.73	1.8	63,400	0.3	7,175	31.6
1923	7,898	70.71	7.6	65,900	1.1	8,378	34.3
1924	8,440	74.21	4.9	74,300	0.7	9,204	35.3
1925	9,065	78.57	5.9	77,100	0.8	10,381	36.2
1926	9,599	81.95	4.3	81,900	0.7	11,173	37.3
1927	9,739	82.10	0.2	85,700	0.2	12,210	38.8
1928	10,234	85.27	3.9	87,500	0.6	13,439	41.2

* Estimated from figures for mutual savings banks.

Conclusions

Prohibition has failed to accomplish the social and economic results which were predicted for it.

The indexes of intemperance, following a decline in war years and the first year of prohibition, show a marked upward trend.

Liquor consumption and intemperance are now at least three-fourths of what they were in 1916, the peak year of the pre-prohibition period, and they are still going up.

Evidence from other countries shows that more progress can be made in temperance reform by regulated sale of liquor rather than by the attempt to impose universal total abstinence by law.

IX

FALLACIES IN PROHIBITION STATISTICS

HERMAN FELDMAN

IN THIS era of prohibition fanaticism, it may not be amiss for me to state that I am neither a Wet nor a Dry nor a teetotaler, and that I will make no discourteous remonstrance if any one wishes to make my Christmas holidays the pleasanter by a proffer of well-tested beverage. I have no emotions on the prohibition question except when I read the research offerings of one side or the other, and then I am alternately a Wet or a Dry, depending upon the kind of statistical material presented. In the present case it is a matter both of pleasure and of regret that I find myself called upon to comment on the data presented by my good friend Gebhart,[1] the most competent student of the subject who has been drawn into a professional relationship to the controversy. Mr. Gebhart's paper is extremely interesting and valuable, and yet it seems to disregard many of the complexities of the subject which should make a careful student less certain about possible deductions from the data.

The present collection of papers has the broad purpose of appraising the place of statistical method in social studies. In the light of this purpose, the first

[1] Professor Feldman's paper was originally prepared as a *discussion* of the paper by Mr. Gebhart (Chapter VIII), both being presented at a *meeting*. While the paper has been revised for the purposes of this book, it has seemed desirable to allow it to remain unchanged in form as a critique of a particular statement by an avowed partisan.—Editor.

point of criticism of prohibition statistics is suggested by Mr. Gebhart's opening assertion: "We need a scientific approach to the study of the liquor problem in this country. Too often protagonists, either for or against prohibition, have presented only those facts which support their position, and have ignored or failed to give proper weight to those facts which are unfavorable."

With this statement one must agree. Nor is the difficulty of which Mr. Gebhart complains, peculiar to prohibition. It is observable in the so-called research conducted in many fields. There are numerous causes for this situation, social and personal. Space need not be taken to recount them in detail. There is one, however, which may well be selected for emphasis because it is fundamental to the future prestige of statistical research—the danger of ruining good research and poisoning the well of information at its source by having statistics gathered by a "National Association for This," or an "American Association Against That."

Any association which assembles a membership of selected mortals who wish something done, and secures funds from them to compile facts to prove something, invariably puts a strain on some of the precious elements of the human mind and of statistical method, namely, balanced judgment and impartiality of purpose. It matters little how competent is the specialist who is hired for the job. We cannot disregard human facts of universal experience. The higher the prestige of the citizens who compose such an association and the greater the dignity and the remuneration of the research specialist employed, the stronger will be the pull on him to please his superiors rather than to play freely with his data. The greater, too, becomes the inertia in a man's critical faculties. The writer has known instances where the situation made the statistician remarkably acute in seeing other people's errors at the same time that it applied a local anaesthetic to that part of his brain which involved the faults of his own data. In time the process grows into the mind to produce a functional disorder not unfamiliar to the psychiatrist. And the finer the statistician employed, the more subtly and intangibly misleading is the research work.

It is my contention, therefore, that the time is passing when any association for or against this or that may engage in statistical research of a controversial character through its normal administrative machinery. If it is sincerely interested in *scientific* research, it should appoint a special research committee on which its worst opponents will have adequate representation, and make that group responsible for the results. The method of the National Bureau of Economic Research may be commended as a tentative model for this new organization of statistical research. The method of the National Association Against the Prohibition Amendment is archaic, though no less archaic than that of similar organizations among the Drys, most of which would do well to emulate the excellence of the material in Mr. Gebhart's comprehensive presentation.

It is beyond the capacity of any ordinary student of the problem to criticize the estimates he gives of present-day consumption of alcoholic beverages. This is the kind of figuring which requires minute study and the use of nice judgment on so many points that the only possible safeguard to the judgment of a careful citizen is to refuse to accept general data from any source that may have been unconsciously motivated in the slightest degree by bias.

An example of the kind of reasoning which suggests a failure to examine all the statistical factors and trends that may play a part in a situation is Mr. Gebhart's statement to the effect that if the net drink bill in 1917 was $1,474,000,000, then: "*Allowing for the increase in population*, we estimate that if there were no prohibition, the probable drink bill today would be $2,500,000,000."

The basic error here is in the assumption that the mere increase in population is a safe guide for determining the probable increase in consumption of any commodity. It totally ignores the enormous increase of prosperity which far outran population and the vastly augmented power of masses of the people to purchase comforts and luxuries. If we assume that the mere increase of population were any index whatever of present-day consumption, the number of passenger cars, which in 1917 was 4,657,340, would be in 1927 not 20,219,224, but about 5,500,000. The purchase of cigarettes, which in 1913 was less than 16,000,000,000, would be not 100,000,000,000, but about 20,000,000,000, and the amount of foreign travel would be a fraction of that which is known to exist.

The writer's contention is, therefore, that in view of the known fact that the consumption of alcoholic beverages closely correlates with the conditions of prosperity of a country, the amount of liquor consumed, if there had been no prohibition, would have shown a sharp upward turn far different from the trend indicated in the decades prior to 1917. How much the difference in trend would have been, cannot be said, though the writer is hoping to prepare some data which will demonstrate the possibilities. But it is safe to say that the Association is faulty in making an assumption which is pleasing to the case, but which totally ignores obvious probabilities.

Likewise, if one is wholly objective about the situation and, therefore, in some doubt as to the interpretation of the figures, one should not use the record of the foreign countries without some interpretation. For example, one cannot cite the reduction in per capita consumption in Great Britain without taking into account the depressed post-war conditions, the high record taxes levied in order to pay war debts, and facts of general poverty and unemployment. The latter, aside from any great trend toward temperance, would cause a reduction in liquor consumption. By ignoring the enormous increase in prosperity in this country, which might have given us a far greater increase in consumption than a mere population trend line would indicate, and at the same time by ignoring the economic conditions which would produce an opposite tendency in England, we might get a striking contrast in statistical terms, but one which would be devoid of scientific validity in respect to the effects of prohibition.

One of the great enigmas of prohibition statistics is the "favorable" character of all indexes in 1920 and the subsequent change for the "worse." There has never appeared any adequate analysis of what there was about that happy year following the war which may be responsible for the exceptionally "favorable" conditions. For example, how much credit may be given to the forced abstinence and improvement in food habits which the war temporarily enforced, and to the educational influences of the marvelous colleges of sanitation which the war camps represented in the case of millions of men under their jurisdiction? In the absence of an attempt to analyze those factors, crude comparisons of numerical data are futile.

Mr. Gebhart refers to the high coefficient of correlation between liquor consumption and alcoholic death rates. This is very interesting, but only those countries may be included which drink the normal alcoholic beverages. When such a correlation is used to estimate the amount of alcoholic consumption in the United States today, all of the statistical technique may be wasted because the character of the liquor has changed. I am not now referring to poisoned liquor, or liquor in which small quantities of poison are to be found. The main point which vitiates the use of correlation methods with respect to consumption in the United States during the past few years is the fact that liquor not prepared under the conditions which prevailed formerly will inevitably produce a much higher death rate.

To make this point clearer it is necessary to discuss with a bit of detail certain aspects in the process of making liquor. The main fact to remember is that in the manufacture of spirits there are, in addition to ethyl alcohol, varying quantities of other substances or by-products formed before, during, or after fermentation, or preëxistent in the raw materials. The nature of these by-products is complex. It is a recognized fact that alcoholic beverages improve with age. In chemical terms, there is little doubt that this improvement is due largely to changes in the character and the relative proportion of these secondary ingredients. It should now be borne in mind that the nature of the physiological effects produced by the consumption of alcoholic beverages depends not merely upon the quantity of the alcoholic content, but on the quantity of the secondary ingredients. An excerpt from an article by an authority on the subject best illustrates the point being made:

> Numerous investigations have been conducted to determine the effects of the various by-products. Thus Dujardin-Beaumetz and Audige found that the pig, whose digestive organs are very similar to those of man, when fed for 30 months with pure alcohol, suffered no ill effects, whereas when similarly treated with imperfectly purified spirit (whether derived from beet, potato or grain) suffered considerably. The same investigators, experimenting on dogs, found that the lethal dose of various alcohols was: ethyl 5 to 6 grammes, per kilo body weight, propyl 3.75 grammes, butyl 1.8 grammes and n-amyl 1.5 grammes. Brunton and F. W. Tunnicliffe demonstrated a poisonous action of furfural upon man and, comparing the after effects upon animals of spirits containing, and freed from, aldehydes, found certain important physiological differences between them.[2]

In the light of these facts it is obvious, therefore, that American alcoholic beverages, which are made largely in haste and under poor conditions, may vary so much in their physiological effect from the beverages used in other countries that there may be little basis for assuming that alcoholic deaths are an index to the total volume of consumption. Instead, such an index may reveal a *qualitative* rather than a *quantitative* difference.

Finally, one must agree with Mr. Gebhart that: "Extravagant claims have been made as to the effect which prohibition has had on increased productivity of labor." At the same time, one may deplore his equally extravagant omission of any mention of the possibility that the abolition of tens of thousands of rotten saloons in the proximity of American industrial

[2] *The Encyclopædia Britannica*, 14th edition, 1929. Vol. 21, pp 245–46.

plants may have had some effect on that productivity. It is too much to believe that the reports that have come from hundreds of employers who are themselves opposed to prohibition can be dismissed with the slight attention he gives to the whole subject.

Mr. Gebhart has not established all of the conclusions that he has drawn; nor does he have a logical right to make them. He has presented some interesting statistical facts, but has he scientifically proved his first sweeping and major contention that "prohibition has failed to accomplish the social and economic results which were predicted for it"?

X

A CRITICAL EXAMINATION OF CERTAIN PROHIBITION STATISTICS

IRVING FISHER

THE purpose of this paper is to examine critically certain of the statistics which have been compiled to ascertain the success of prohibition. In particular, it is directed to the consideration of the paper of Mr. John C. Gebhart, "Prohibition: Statistical Studies of Enforcement and Social Effects," presented as Chapter VIII of this volume. With Mr. Gebhart's first statement that we need a scientific approach to the study of prohibition, I most heartily agree. This approach has been the basis of my attempts to gather whatever available data there may be which indicate whether or not prohibition is succeeding, and if not, to what extent it is failing.

I wish to emphasize at the outset that in all phases of my study of the prohibition question, I have striven to maintain a strictly scientific attitude. In my experience I have had many opportunities to observe the verification of the aphorism that even the propositions of Euclid would be vigorously challenged if they were thrown into politics. When an important national issue becomes a football of political discussion, the fairmindedness of everyone is challenged, no matter how persistently he tries to find the truth. There have been some who have tried to discredit my efforts, either by concluding or by pretending to conclude that, because I had reached results different from theirs, I must be prejudiced.

All that I have tried to do, in the attention which I have given to the problem, is to appraise as dispassionately as possible what success prohibition has had. The mere claim of unprejudice, however, is sometimes taken as proof of prejudice. Nevertheless, I can offer evidence that I have tried to follow the facts.

When, after having published a book, *Prohibition at its Worst*, I found through criticisms—by both friends and enemies of prohibition—that I had made mistakes, I freely admitted those mistakes in a succeeding book, *Prohibition Still at its Worst*. In particular, I may refer to a correction of one of these mistakes which I made in the light of later acquired information—a correction which I am rather surprised that Mr. Gebhart does not mention.

Mr. Gebhart writes that I am "able to show"—a phrase which perhaps attracts my attention more than that of others—that, according to the statistics of arrests for intoxication, there has been a falling off in drunkenness.[1] This apparently implies, whether intentionally or not, that I have deliberately attempted to get a preconceived conclusion out of these data. The truth is that I have had no faith in statistics of arrests for drunkenness as an index of prohibition enforcement; nor, in fact, in any of the indexes which, according to Mr. Gebhart, "are accepted by both Wet and Dry partisans." I think that all these indexes are very questionable. I certainly do not regard arrests for drunkenness as having the reliability which Mr. Gebhart implies.

I have probably been misunderstood with reference to my attitude toward statistics of arrests because of the attention given to these data in my first book.

[1] Cf. Gebhart, p. 117.

In this instance, extended consideration of this index was required because of the importance placed upon it by the Moderation League, whose figures Mr. Gebhart uses. The Moderation League, in my opinion, had drawn an unwarranted conclusion from the statistics of arrests. After a careful study of the figures, I concluded that the changed and changing policy of the police must be allowed for in using these data. If I understand Mr. Gebhart correctly, he agrees with me on this point; yet, he has not attempted to make any correction for this factor, and lacking this correction, the figures for arrests are almost worthless.

When I wrote my first book, I had before me the figures of Mr. Robert E. Corradini, of the World League Against Alcoholism, who had attempted to discover the ratios between arrests and actual drunkenness before and after prohibition. His estimates indicate that before prohibition forty percent of the actual drunkenness was shown by the arrests and after prohibition ninety percent. These figures were conscientiously compiled on the basis of police department records throughout the country. When I used the figures, they were criticized—with some justice, I thought—particularly because I had applied them indiscriminately to each individual city. This, I now grant, should not have been done, because of the wide difference in the changes of police policy in New York, Los Angeles, Chicago, and other cities.

Having made this mistake, I decided to review the whole subject. Questionnaires were sent to about one thousand chiefs of police throughout the nation asking for verification, or the opposite, of Mr. Corradini's estimates. There is no secret about this questionnaire, and I shall gladly furnish it to anyone interested. The tabulated results were not published because I felt that more attention had already been given the subject than it really deserved.

I did, however, make every effort to derive from the questionnaires all of the information they contained which would shed light on Mr. Corradini's estimates of forty and ninety percent. Instead of tabulating the results myself, I had this work done by an expert statistician (whose prejudice, if any, is against the Eighteenth Amendment). The results of this tabulation I accepted even though they were considerably more unfavorable to the prohibition cause than Mr. Corradini's figures. This questionnaire showed that, instead of the forty percent, fifty-five percent were arrested before prohibition; and instead of ninety percent, seventy percent were arrested after prohibition.

When I corrected the figures on arrests, I found that I was "able to show" (if I had tried to show anything) a much larger reduction by continuing to use the older figures of Mr. Corradini. Had I done this, no one would have been the wiser. But instead, I employed the revised ratios, though the results I got thereby and published in my second book were not as helpful to the prohibition cause as those which had been published in the first book. I made the change because I thought it was in the interest of truth. I could give many other instances of this same effort on my part to get at the exact facts.

Mr. Gebhart speaks of the effects of the use, disuse, and revived use of the "Golden Rule" policy upon the statistics of arrests for intoxication. I think that the most valuable part of his paper is that which draws attention to the effects of this policy on the data of arrests for Detroit. As his chart with respect to that city shows,[2] if account is not taken of the change in this policy, the conclusion can be drawn either that there was a great increase in arrests or that there was an improvement in enforcement at the end of the period. The conclusion may be mistaken in either case.

I understand that, except for the last period when the "Golden Rule" was reintroduced, there is in almost every other city in the United States a situation similar to that depicted by the Detroit curve. There has been an abandonment of the "Golden Rule" policy. In the articles written by Mrs. Willebrandt, which appeared in the newspapers, the fact is noted that the abandonment of the "Golden Rule" treatment in Los Angeles and elsewhere represented a change in policy which gives a false and exaggerated picture of the increase of arrests.

There was, in my opinion, a strict and sudden enforcement of the prohibition laws immediately following the precursors of constitutional prohibition, i.e., first, the war-time restrictions on the liquor traffic, and second, the war-time prohibition. There has been some effort to discredit this view and to say that these early results cannot be credited to prohibition since real prohibition began in 1920. But the fact is that we did have war-time restriction in 1917 and 1918, and war-time prohibition in 1919, and each of these was practically equivalent to prohibition as at present. It seems probable that the curves depicting the success of enforcement in 1917, 1918, and 1919 would not have been very different had constitutional prohibition been in effect.

The success of prohibition at the beginning is evidenced in the tremendous drop in the number of arrests for drunkenness and of deaths from alcoholism

[2] Cf. p. 121.

after war-time restrictions and war-time prohibition. Since then, there has been a retrogression in the success of enforcement. The entire experience is depicted by the curve presented by Mr. Gebhart. This curve has the appearance of a "fishhook," and is characteristic of most prohibition statistics. There is usually a sudden dip downward just after the passage of prohibition statutes, indicating a strict enforcement. The low point generally comes in 1920, the first year of constitutional prohibition. There then follows a period of relaxation in enforcement, at first very rapid and then showing a tendency to flatten out on a level somewhere between the original pre-prohibition level and the lowest level reached.

I am inclined to think that we are now in a stage properly called prohibition at its worst, that this curve which is tending to flatten out will, if we persist in our efforts to make prohibition a greater success, turn downward again. I doubt whether this curve will ever reach again the low level of 1920, but I think that even at its present level prohibition has been of great value.

The data of deaths from alcoholism do not have the value which Mr. Gebhart seems to attribute to them. In connection with the use of these data, there is a consideration quite similar to that involved in using the statistics of arrests. While in this latter instance changes in police severity tend to vitiate the figures for the purpose for which they are used, in the former case allowance must be made for changes in the accuracy of reporting the causes of death. When, as has been true at certain times, it has seemed more or less disgraceful to attribute a death to alcoholism, deaths have often been recorded as occurring from other causes. So there is a ratio or coefficient of the faithfulness in reporting the causes of death which must be taken into account. As far as I know, no effort has as yet been made to ascertain this coefficient.

There is, moreover, another factor which should be considered when deaths from alcoholism are employed as an index of prohibition enforcement, or of liquor consumption. Acute alcoholism is largely due to the sudden upsetting of physiological balance. It occurs before Christmas and on New Year's Eve, when there is a large amount of alcohol imbibed during a short time. This type of drinking does not necessarily indicate that there is a large amount of alcohol imbibed on the average. One of the things which prohibition has brought about is the inability of people to obtain alcohol continuously every day of the week. It has therefore led to periodic and sudden sprees. The drinking in schools and colleges, for example, is not as great now as it was before prohibition. When it does occur, it is more extreme.

On account of this irregularity in drinking, an increased number of deaths from alcoholism would be expected. But the increased number of deaths does not indicate increased drinking on the average. I am inclined to think, although I have not absolutely closed my mind on the point, that the statistics of deaths from sclerosis of the liver are a better indication of the total consumption of alcohol than deaths from alcoholism, because the latter deaths result from periodic sprees while the former result from the cumulative effects of continuous drinking.

The case of New York, as an example of the failure of prohibition, seems to me questionable because there is no real prohibition enforcement there. The state has repudiated the Mullen-Gage Act by which it was to assume the same responsibilities under the prohibition statutes as the other states. Similarly, in Maryland, there has never been an enforcement law, and consequently a greater degree of failure exists there. In these two states, the evidence may more properly be taken to indicate what may be expected where prohibition is lacking than what is to be obtained from prohibition itself.

I may note the well-advised omission of Chicago from Mr. Gebhart's compilation of arrests because the Chicago figures include other arrests than for drunkenness, namely, those for "disorderly conduct." No mention, however, was made by Mr. Gebhart of Indianapolis, and I do not know whether he includes that city in his figures. Its exclusion is even more essential than that of Chicago. No one knows the source of the figures for Indianapolis given by the Moderation League. A representative whom I sent there learned that the Police Department and every other department that could possibly have supplied them knew nothing about their source, and repudiated them. They seem to have come out of the air, but they weigh down the total for the United States in a most amazing way.

The real problem with which we are concerned is whether or not the consumption of alcohol has been reduced through prohibition, and how much. Mr. Gebhart's estimates of the amount of liquor consumed seems chiefly valuable as discrediting the estimate made by Mr. Hugh Fox, Secretary of the United States Brewers' Association. Mr. Fox had an astonishing series of figures by which he was "able to show" that there was an actual increase in the consumption of liquor. Mr. Gebhart, however, was not "able to show" this. His figures indicate that the consumption of liquor is only three-quarters of what it was before prohibition.

Mr. Fox had a figure for the consumption of spirits of 251,000,000 gallons as against Mr. Gebhart's estimate of 181.000.000 gallons. The latter figure is about thirteen times as great as the estimate of 14,000,000 gallons by Mr. Corradini, whose calculations were thorough, conscientious and careful. The difference is very largely due to the estimated consumption of corn sugar employed in the manufacture of spirits.

Mr. Gebhart estimates the consumption of beer at 440,000.000 gallons as against Mr. Corradini's estimate of 325,000,000 gallons. In the case of wines, Mr. Gebhart's estimate is 150,000,000 gallons as against Mr. Corradini's estimate of 48,000,000 gallons —less than one-third as much.

As Mr. Gebhart himself admits, no one knows how much liquor is being consumed in the United States, but even using his own figures it must be far less than before prohibition; while, if we use Mr. Corradini's figures, present consumption would be only a small fraction of the pre-prohibition quantity.

It must not be forgotten that the alcohol problem is really insoluble during the present generation. Habits which have been handed down by tradition for many generations cannot, as Mr. Gebhart says, be suddenly changed by law. Nevertheless, experience with state prohibition in Kansas, and in other states that have had prohibition in force for a long time, indicate that, by keeping everlastingly at it, and by waiting until the deaths from alcoholism have overtaken some of the people who will not obey the law, we shall ultimately achieve a greater degree of success.

When people say prohibition is a failure, I answer,

"Yes, it is a *partial* failure." But every other solution of the liquor problem which has been tried has been a partial failure. I am satisfied that prohibition, especially in the United States, has succeeded more fully, and is much more likely to succeed, than any other method.

Consideration should be given in this connection to the data for Canada, which, as used by Mr. Gebhart, consist of the official estimates of the consumption of alcohol. These estimates do not take into account the consumption of alcohol resulting from bootlegging. After sending a representative to Canada to study the situation on the ground in an attempt to get at the facts, I am convinced that Canada is finding its experiments with the liquor problem partial failures.

The same inconclusiveness of results will follow every method of dealing with this age-old problem. One hundred percent success has not yet been reached with the Canadian, Scandinavian, or with any other system. As to the United States, we should keep at the experiment which we are now making until a substantial agreement is attained as to its ultimate result.

I believe in prohibition on the grounds of expediency rather than of principle. I believe in giving the experiment a fair trial. Originally, I had been opposed to prohibition, but became convinced of its value on the basis of experience with state prohibition. Later, when, during the war, I was chairman of the committee on alcohol of the Council of National Defense, the committee unanimously recommended that war-time prohibition should be introduced as an experiment. This recommendation called for the continuation of prohibition for one year after the war, at the end of which time the public could consider the question of its continuance in an atmosphere unaffected by war or other irrelevant issues.

What we got, however, is something quite different from what was contemplated by the recommendation of this committee. Instead of a flexible prohibition experiment, we adopted constitutional prohibition. Under these conditions, the questions are whether we shall accept it and strive to make it a success, whether we shall repeal it and substitute something else, or whether we shall attempt to modify it. Of these three options, I personally choose the first.

Although I believe that constitutional prohibition came prematurely, I have a considerable degree of confidence that, if we follow the lead of the President, he may, during his term of office, make the present experiment much more of a success than it has been. I have been disappointed with the record of the past, as have others, and yet, even Mr. Gebhart shows that the experiment has been at least a partial success.

XI

STATISTICAL STUDIES OF SOCIAL ATTITUDES AND PUBLIC OPINION

STUART A. RICE

STATISTICAL analysis is only employed where a plurality of individual items or units is involved. It is commonplace to say that the first task of the statistician is to define his units.[1] It is not so often pointed out that units are frequently arbitrary, and that they may depend upon a prior conceptual, analytic, or philosophical process. A given field of investigation may be so formulated, on the one hand, that a plurality of units does not appear. This formulation will preclude the use of statistics, or at least direct attention away from the possibilities of statistical statement. On the other hand, the formulation may be such as to render the statistical mode of attack inevitable and imperative, or at least to call attention to its possibilities and advantages.

Consider an illustration which will anticipate, by a few pages, the later discussion: In some Pennsylvania communities it is believed that if the shavings resulting from the manufacture of a coffin are not swept up and placed therein prior to a burial their presence among the survivors will bring further death or misfortune. If, without further instructions, a teacher were to ask a class of graduate students, "How would you proceed to investigate such a superstition as this?" he would

[1] Some interesting individual problems of definition are set forth in the papers by Hugh Carter, Donald Young, and Clifford Kirkpatrick in this volume.

probably receive two types of replies: a portion of the class would propose to inquire into the historical circumstances under which the superstition arose and under which it was evolved and diffused; another portion of the class would suggest attempts to explain why the superstition should gain credence in human minds. Neither group would be likely to propose statistical modes of inquiry, because the historical and psychological approaches are suggested when the belief is labeled with the concept *superstition.* Suppose, however, that the class is asked, "How would you proceed to investigate such a superstitious attitude as this?" Without any restatement of the data and by inserting the single word "attitude," the teacher has changed the connotation of the question and suggested a new type of inquiry. The word "superstition" seems to refer to a discrete psychological entity, one which is either held or not held by individuals. The word "attitude" evokes a concept of something variable, to be examined statistically. The extent to which the two questions actually evoked these two concepts, respectively, would, of course, depend upon the extent to which the suggestions which the terms contain were uncritically accepted in the minds of the students. For instance, one might *say* "superstition" and the student of independent type might *think:* "attitude—hence, variable."

The process of conceptual formulation has been presented so far as one which precedes and directs the choice of further means of inquiry. But the sequence may be reversed. The desire to proceed quantitatively, for example, may bring about a conceptual reformulation of the subject matter. This, we have been told by Mitchell, is something which has actually taken place in the recent development of the science of economics.[2] A similar trend is discernible in political science.

This discipline, like history, has until recently been largely concerned with legal formulations; and with unique, non-repetitive situations. When James Bryce discussed "modern democracies" he did indeed utilize what Bernard and Kirkpatrick call "the informal statistical method." [3] But it was more usual for political scientists to discuss single states, and to deal with formal aspects of their constitutional organization and activities. One cannot use statistics readily in explaining the constitutional limitations imposed upon the authority of the Federal Government, or in describing the history of diplomatic relations between the United States and Great Britain. But when the general notion of the State becomes less absolute and more functional, when sovereignty comes increasingly to be regarded as pluralistic, when formalism is replaced by realism as a premise in the observation of events, then statistical method begins to have a rôle to play. Such actual changes as these in the developing concepts of political science have resulted in the use of statistical methods. At the same time, they have been a *result of* the prestige attached to statistics, especially in the sister subject of economics.

Florence, like Catlin,[4] does not even limit the scope of political inquiry to the State. In his recent important book, Florence sets out "to provide for statistical politics the analytic framework, and the

[2] Mitchell, Wesley C., "Quantitative Analysis in Economic Theory," *American Economic Review*, XV, 1925, pp. 1–12.

[3] Cf. the paper by Clifford Kirkpatrick in the present volume.

[4] Florence, P. Sargant, *The Statistical Method in Economics and Political Science*. New York: Harcourt, Brace & Co., 1929. Catlin, George E. G., *The Science and Method of Politics*. New York: A. A. Knopf, 1927.

apparatus of thought" that have already been so largely achieved in economics. Hence, political science "is interested in the acts or behaviour of men toward men, their mutual interrelations and reciprocal contacts; the orders, punishments, votes, verdicts, appointments, dismissals, passing transitively from men to men, and the meetings and discussions between men, that form part of their relations of ruling, manning and sharing of work" (364). So viewed, statistical politics becomes not so much a possibility as a necessity, and the author is able to itemize in considerable detail the statistical inquiries which are required to fill in the frame.

When we come to the subject of attitudes and opinions, then, the first task appears to be an examination of the concepts, and attendant definitions, upon which statistical treatment depends. The process of conceptual formulation has gone on rapidly during the past half decade. It must be admitted that there is still lacking that extent of agreement among investigators which gives assurance that all are talking about the same entities when the same terms are used. (However, this is still quite generally the situation in social science.) Two developments of thought which seem to have laid the basis for the present interest in attitude measurement will be indicated at this point.

The first of these developments has been pointed out by Franklin Fearing in a survey of the subject, as yet unpublished.[5] He notes that present interest in the study of attitudes represents a reaction from the stimulus-response psychology of a few years ago. That is, it represents renewed direction of attention

[5] This will appear in the "Case Book" on the methods of social science, now in preparation for the Social Science Research Council, under the editorship of the present writer.

inward, toward directing motivations in the organism itself, and away from those factors in the external environment that provide the stimulus. By way of definition, Fearing says: "All those factors involved in the readiness or preparedness of the organism to respond are referred to under the terms which we have used, i.e., determining tendencies, mental set and attitude. These include all the processes preceding and determining the motor side of the act, not excepting those processes which antedate the stimulus itself. The latter becomes merely a releasing mechanism."

This might seem to remove investigation entirely from the sociological to the psychological level. It would do so except for the second of the developments to which reference has been made. This appears when the concept of attitude, as just set forth by Fearing, is compared with the concept of instinct, recently so prevalent. In once more directing his attention toward the drives within the individual, the investigator has not returned to the instinct hypothesis. He has taken with him the sociological concept of personality as a resultant of social and cultural *experience*. Instincts and attitudes would otherwise be identities concealed under different names. Instincts are conceived as inborn; attitudes as composites of inborn drives and experience. Thus, F. H. Allport in a publication of 1924 alludes to "the motor set thus built up by suggestion" (i.e., by experience) which "we may call an *attitude*." [6] The same distinction is found in a more recent "temporary" definition by another psychologist, G. W. Allport. He refers to attitude as "a disposition to act which is built up by the integration of numerous

[6] Allport, Floyd Henry, *Social Psychology*. Boston: Houghton Mifflin Company, 1924.

specific responses of a similar type, but which exists as a general neural 'set,' and when activated by a specific stimulus results in behavior that is more obviously a function of the disposition than of the activating stimulus. The important thing to note about this definition is that it considers attitudes as broad, generic (not simple and specific) determinants of behavior." [7] Attitudes are "built up by . . . responses" not given in the germ plasm at the start of the individual's life.

The consequences for our subject of this distinction between instincts and attitudes are important. The concept of instinct puts an emphasis upon the discrete character of the entity. The concept of attitude emphasizes its variable character. The distinction is not, however, logic-tight. C. Kirkpatrick and J. W. Woodard have called my attention to investigations posited upon the assumption that instincts are variable, and a suggestion of similar import has been made by W. F. Ogburn in his book, *Social Change.* The most that can be contended is that the concept of instinct does not seem to *favor* such investigations, while the concept of attitude does favor them. For if we start with the concept of specific instincts, as did William McDougall, our major problem becomes one of identifying these instincts. We are interested in whether or not certain drives are to be found within us, rather than in variations in the intensity of these drives. But if we conceive of attitudes as resultants, in part, of experience, we necessarily have our attention focused upon their variability. Since the experiences of people vary, so must also their attitudes, if the latter grow out of the former. When con-

[7] Allport, Gordon W., "The Composition of Political Attitudes," *American Journal of Sociology*, XXXV, Sept., 1929, pp. 220–38.

fronted with a given stimulus, which more or less resembles a greater or larger number of other stimuli hitherto responded to, the individual attitudes evoked in a number of persons must be conceived as forming, potentially at least, a continuum along a scale, when evaluated with respect to any quality such as direction or intensity.

Two other theoretical considerations are relevant to the discussion at this point, before taking up the topic of *opinion.* Some years ago Faris raised the question whether, in discussing instincts, we were dealing with data or hypotheses.[8] It is sometimes suggested that the same question is now timely with respect to attitudes. Are attitudes hypotheses rather than data? The question seems debatable, and perhaps the answer turns upon definition. At the same time, an affirmative answer would not be especially damning. The concept of attitude, hypothesis though it may be, provides a methodological postulate upon which to base interesting and useful classifications of human behavior. Positing the attitude, we seek to find behavior which we may attribute to it. Measurements of the behavior then provide us with indexes of the attitude, which is thereby defined in terms of the behavior. It is in this sense that it seems to the writer to be permissible to say that attitudes can be measured. Moreover, it seems a useful assertion to make, provided the conceptual or hypothetical nature of the terms be not forgotten. It is more practicable to call a man "a conservative" than to describe on each occasion the behavioristic class into which we would put him on the basis of what he says and does.

[8] Faris, Ellsworth, "Are Instincts Data or Hypotheses?" *American Journal of Sociology*, XXVII, pp. 184–96.

The second theoretical consideration concerns the possibility that attitudes (and opinions) might be studied statistically as they are found at different times or under different circumstances within a single individual. This suggestion is implicit in Thurstone's formulation of the "law of comparative judgment." [9] No instructor, for example, can regrade a set of examination papers with his first marks obscured and award precisely the same grades on both occasions. In a series of decisions affecting the rights and privileges of labor, we might expect an honest and seasoned jurist to be at times more favorable and at times less favorable to labor's cause. The curve expressing the distribution of such variations might, of course, be distorted from the familiar bell-shaped form because of the cumulative influence of habit or of varying trends in the learning process. In this case, perhaps the concept of a *time series* of attitude expressions, to be measured by a trend line, would be more applicable.[10] However, the present paper is concerned, not with variations of attitude *within* individuals, but with variations of attitude *among* individuals. This lightens the task of establishing units that may be statistically enumerated, for while the psychological states with which we are concerned may remain obscure, the human "carriers" of these psychological states are distinct. As Thurstone has pointed out, the unit may consist either of a single average expression of attitude upon a given topic by a single person; or it may consist of any one among a plural

[9] Thurstone, L. L., "The Law of Comparative Judgment," *Psychological Review*, July, 1927.

[10] These suggestions seem to indicate the possibility of an introduction into the field of jurisprudence of a type of statistical methodology even more novel than the developments discussed by Professor L. C. Marshall, Chapter VII.

number of expressions of attitude upon the topic by *a single person.* That is, there may be exactly as many attitudinal units to enumerate as there are persons; or there may be more attitudinal units than persons. In either case, the process of identifying and counting the units is vastly simplified by their identification with persons.

The argument with respect to attitudes up to this point may now be summarized: We are dealing with psychological entities, "real" or "hypothetical," conceived as variable among individual persons, the essential nature of which if "real" we know little about, but which are brought to our sense perceptions by behavioristic evidences [11] which may be distinguished into units, classified, and enumerated. Behavioristic evidences have not yet been discussed. The topic brings us to a consideration of *opinion* and its relation to attitude.

In the first noteworthy American effort to measure the distribution of opinions and to indicate their specific relationships to attitude, Allport and Hartman [12] do not develop clearly the general dependence between the two concepts. At the Round Table on Political Statistics of the Second National Conference on the Science of Politics,[13] protracted discussion produced agreement on three points only concerning a definition of opinion: (1) "It need not be the result of a rational process; (2) it need not include an awareness of choice; and (3) it must be sufficiently clear or

[11] To please the introspectionist we might add: *in the case of other persons.*

[12] Allport, Floyd H., and Hartman, D. A., "The Measurement and Motivation of Atypical Opinion in a Certain Group," *American Political Science Review*, XIX, Nov., 1925, pp. 735–60.

[13] Chicago, Sept. 8–12, 1924. See reports in *American Political Science Review*, XIX, Feb., 1925, pp. 104–62.

definite to create a disposition to act upon it under favorable circumstances." Further, "On the question when is opinion public, the round table was unable to come to a definite conclusion." It will be noted that this attempt to define opinion produced a concept strongly resembling that of attitude, as previously defined. In connection with efforts to measure various political phenomena, the writer expressed a preference "to avoid the use of the term opinion and to use instead the word *attitude*, as indicating a disposition or set toward behavior without reference to the degree of rationality that may be present in connection with it. . . . 'Opinion' and 'public opinion' seem . . . to connote too much of the rational and conscious elements in the actual motivation." He therefore proceeded to "use the word attitude in a somewhat inclusive sense, without endeavoring to determine in any particular case whether the word opinion might be preferable." [14]

A formal attempt to relate the concepts of opinion and attitude for methodological purposes has been made by Thurstone.[15] "The concept 'attitude,'" he says, "will be used here to denote the sum total of a man's inclinations and feelings, prejudice or bias, preconceived notions, ideas, fears, threats, and convictions about any specified topic. . . . The concept 'opinion' will here mean a verbal expression of atti-

[14] Rice, Stuart A., *Quantitative Methods in Politics.* New York: Alfred A. Knopf, 1928, pp. 51–52. In criticizing the manuscript of this paper, Professor Gordon W. Allport objects to the writer's identification of "opinion" with "the rational and conscious elements" in motivation.

[15] The discussion of Thurstone's work, unless otherwise noted, will be based upon the monograph, Thurstone, L. L., and Chave, E. J., *The Measurement of Attitude, a Psychophysical Method and Some Experiments with a Scale for Measuring Attitude toward the Church.* Chicago: University of Chicago Press, 1929.

tude." This distinction has an obvious methodological advantage. Opinions are accessible. Either the opinion itself or the person holding it may be treated as a unit, and classification may then follow, thereby permitting statistical enumeration and analysis. If opinion is regarded as an index of attitude, which Thurstone regards as below the verbal though not necessarily below the reflective level, then attitudes gain a reflected objectivity and precision from the opinions which represent them.

A criticism of Thurstone's concept of the relationship between attitude and opinion will be withheld for a little, in order to proceed directly to an examination of the essentials of his method. His procedure is the latest attempt, as it is the most complete, rigorous, and rational, to employ statistical methods in the study of attitudes. It will serve in this discussion as the best example of an important and standard type of inquiry in this field.

Thurstone's most important achievement is his method of constructing a rational scale of values, to which any number of relevant and unambiguous propositions falling in the linear dimension of the scale may be related, each with its own scale value. This has not hitherto been accomplished in attitude measurements. The so-called scales established by predecessors have merely consisted of an arrangement of statements in rank order. The intrusion into the series of new statements, in this latter case, would have extended the scale by a corresponding number of points, and thereby have disrupted all of the so-called values with the corresponding frequencies, previously located upon it. As Thurstone points out, the numbers of endorsements received for the various statements on such a rank-order scale do not constitute a frequency distribution in a statistical sense. In Thurstone's scale, on the contrary, propositions may be added or taken away without affecting the scale itself, which is therefore independent of *particular* statements.

The technique employed in Thurstone's method represents, in a broad sense, an endeavor to apply in the fields of attitude and opinion principles developed by psychophysicists. In essence, the method consists of a determination of the intervals which appear equidistant in the average opinion of several hundred judges. A large number of statements, which are intended to include extreme positions at both ends of a linear continuum of opinion, together with intermediate positions, are sorted by each of the judges into eleven classes. These classes appear to each judge to be equally spaced between the two extremes. Each statement will have as many ratings between one and eleven as there are judges. The accumulative proportions of the ratings received by each statement in each of the eleven ordered positions are calculated. These proportions are plotted on rectangular axes, in which the eleven scale positions appear as abscissas and the accumulative proportions as ordinates. The resulting accumulative curve for each proposition, similar to the familiar ogive, is smoothed. The point on the X axis where the curve shows an interpolated accumulative total of fifty percent of the cases is taken as the interpolated scale value of the statement. This scale value is in terms of the eleven equidistant positions into which the base line has been arbitrarily divided. The scale value thus appears as a median valuation by the judges, graphically determined.

The interquartile range, or Q-value, is taken as a measure of ambiguity. "If a statement is very ambiguous, the different readers will place it over a wider range on the scale and the Q-value will be correspondingly high." It may then be rejected in the selection of statements to compose the final scale which is to be used in measuring the attitudes of a group of subjects. Thus, a statement is included in the scale only when the judges are in relatively close agreement as to the position which it should occupy with reference to other statements, and with reference to the extremes of attitude.

The objective criterion of ambiguity which is afforded by the Q-value somewhat mitigates, but does not wholly overcome, what seems to the writer to be an inconsistency between two of the assumptions underlying Thurstone's method. This concerns the relationship of opinions and attitudes. When judges are asked to sort statements into classes, representing equidistant positions along a linear scale, they are asked to make judgments *concerning* attitudes, not to express their *own* attitudes. The judgment expressed may presumably be called an opinion. For example, among a series of statements it is one's opinion, as a judge, that statement A is a reflection of a more pacifistic attitude than is statement B. The judge's own attitude may be quite militaristic, but he is asked to express opinions concerning the comparative pacifism exhibited by statements at the other end, and at all points, of the scale. The task given him assumes that he is capable of expressing a rational opinion concerning the opinions upon the issue in question, independently of his own attitude toward the latter. All of this is in his capacity as a judge. But suppose his own attitudes are then tested by means of the scale that he has helped to construct. It is now assumed that his opinions are no longer independent of his attitudes, but, on the contrary, that the former are indexes or reflections of the latter. In the first case, his opinions *must* be rational; in the second case, they *must not* be rational. It should be clearly noted, however, that this criticism depends upon the legitimacy of referring to *judgments* as *opinions*.

What is the possibility that the acceptance or rejection by a subject of a statement upon the completed scale may represent a rational judgment concerning the truth or falsity of the statement made? It would seem to exist. If so, the validity of the statement as an index of attitude is destroyed or impaired. If I accept the statement that H_2 and O when combined will produce the chemical combination known as water, this does not express an attitude. It is a statement of fact. Might it not be considered a statement of opinion, if there were a degree of uncertainty upon the matter in my mind? If I concur in the proposition that the corn borer is threatening the production of corn in the United States, I might be credited with a factual judgment if I were an agricultural expert. But since I am not, and if I were called upon to support the proposition with evidence, would I not reply that I merely expressed an opinion? My evidence is a bit hazy. My assent to the proposition does not reflect an attitude toward the corn borer, for I may be either a bull or a bear in the corn market. But if I concur in the proposition that "the organized church is an enemy of science and truth," [16] this is taken as evidence of an attitude (having a scale value of 10.7, or a position hostile to the church) in a scale between hostility and favor

[16] Statement 72 in a schedule for the "Experimental Study of Attitude Toward the Church," in *The Measurement of Attitude*, by L. L. Thurstone and E. J. Chave, p. 62.

toward the church. Where is the distinction to be drawn between factual-judgment opinions and attitude-representation opinions? The objective criterion of ambiguity does at least eliminate statements upon which the judges cannot agree closely as to a specific relationship between the opinion and an assumed position on the attitude scale.

Another criticism concerns a subtle ambiguity in the use of the terms "favorable" and "unfavorable" in describing the respective halves of the attitude scale on either side of the neutral point. These terms are not used by Thurstone, but have been employed by others who follow his general procedure. It is possible to think of attitudes in terms of the effect they will have upon the person, policy, or situation toward which they are directed. This is not always easy to distinguish from what may be paradoxically called the intent of the motor set within the subject himself. For example, if I am serving as a judge in the construction of a scale designed to measure the attitudes of white people toward persons of color, the effect upon the colored man may serve as my criterion in distributing the statements into the various classes. In the case of propositions A and B, I may reflect as follows: Proposition A, if generally held by white people, would result in benefit to the Negroes. Proposition B, if generally held, would result in less benefit. Hence, proposition A is more favorable to the Negro than propostion B and should be placed at a higher (or lower) position on the scale. The arrangement of propositions, then, is determined by the state of the object rather than by the state of the subject.

To be specific, let us suppose that the judge is a psychiatrist. He is asked to classify statements concerning the Negro into eleven classes; from Class One, the least favorable, to Class Eleven, the most favorable. Suppose he is confronted with the statement: "I regard the Negro as I do myself, and will share with him whatever I may possess at all times." Is this statement favorable or unfavorable to the Negro? The psychiatrist may regard this as definitely unfavorable, because he thinks that the Negro should be encouraged to develop greater independence than such an attitude by whites would permit. Or, suppose the statement to be: "I am impatient of the Negro's dependence upon sympathetic white men. Force him to work and take care of himself." "Ah!" the psychiatrist might rationally say to himself, "that is just what the Negro needs to throw off the psychological heritage of slavery. That is a favorable attitude!" And would he not be right, assuming the correctness of his premises? The attitude would be favorable toward, i.e., result in benefit to, the colored man. But the psychiatrist would not be building a scale of attitudes, as generally understood, and in the process of scale-building the order of the two propositions cited would probably be reversed. By attitudes we intend to refer to, and should employ terms which express, the disposition of the subject rather than the favorable or unfavorable effect upon the object. Perhaps we might say "favorably disposed" and "unfavorably disposed," as indicating a motivation to favor or the reverse, quite apart from any appraisal or calculation of the probable outcome. In his scale of attitude toward the church, Thurstone uses the terms "*appreciation* of" and "*depreciation* of." These seem unexceptionable.

It may be added, without going into further particulars, that Thurstone has exhibited great skill and care in developing checks and measures of reliability for his calculations at all points.

There is still another conceptual problem involved in the composition of the board of judges. At least three possible types of selection may be considered. In the first, the judges are presumed to be *experts* upon the issues. A second mode of selection calls for a widely selected sample of persons from the general population. A third calls for a sample from the group or class, the attitudes of whose members are later to be tested. Certain difficulties attend each of these modes of selection.

In studies preceding Thurstone's, the persons who have arranged the order of the propositions composing the so-called scale have been presumed to be experts. as, for example, teaching colleagues of the investigator, supposedly familiar with the subject matter contained in the test. Such judges are accustomed to rational reflection upon such issues, and, in particular, upon the immediate issue. Hence, they will tend to classify attitudes as favorable or unfavorable, rather than as favorably disposed or unfavorably disposed, in the sense of the distinction drawn above. The very possession of expertness, then, may be viewed as tending to render the judge incapable of arranging propositions in such a manner that they will seem to non-experts to be in order and to have linear differences equal.

An analogous difficulty applies to the selection of judges who will constitute a sample of the general population. It may be contended that not only the "equal-seeming interval" but the order of propositions itself may differ considerably among different groups, on the average, as a result of differences in cultural background. An opinion characteristically regarded as favorable, or favorably disposed, to the Negro in the South might be characteristically regarded as unfavorable, or unfavorably disposed, in the North, and vice versa.[17]

The third of the three choices proposes the selection for judges of a sample from the same class or group or segment of the population as that to which the scale is to be applied in the subsequent test. It shares another difficulty with the second choice, just examined: Will persons in any sample made up of non-experts have sufficient understanding and command of abstractions, and particularly of language, to verbalize their attitudes? That is, will the average judge in the sample be intellectually capable of making the discriminations called for by his instructions and detaching his own attitudinal bias from the discriminal process? Is the average person competent to perform this dual and rather dexterous task? Perhaps the answer is that these questions apply with much the same force to the subjects to whom the scale is applied. Perhaps the attitude scale, like other scales employed in science, is valid only within the middle ranges of its phenomena—in other words, among persons whose intellectuality and knowledge are neither too great nor too small.

There is still another disappointment in store if the third mode of obtaining judges is employed. One of the major objectives of the method is to compare the attitudes of different groups. If each group requires a measuring scale built up for that group alone, direct comparisons between groups as to average tendencies become impossible, although comparative variability might still be determined.

[17] The writer is indebted to Professor Donald Young for this point, and for instances which seem to illustrate it.

Unsatisfactory as is this third among the choices, because of the limitations that it seems to place upon the use of the scales, it seems to involve less contradiction than do the first and the second. If so, we must be contented to recognize that a scale devised with the aid of judges in Boston, for example, when applied in Atlanta, would interpret attitudes solely in terms of, or relative to, a Bostonian culture pattern.

It may be comforting to recall that similar considerations have often been overlooked in intelligence measurements. Even the familiar scales of physical measurement are regarded by physicists as relative, and as inapplicable when the physical constants change appreciably. Thurstone has at least made a beginning, and it seems to the writer a first beginning, in the exact and rational application of psychophysical theory to the construction of an opinion scale, with resulting occasion for the use of statistical methodology.

Statements of opinion, however, are to be regarded as but one among various forms of expression of attitude. Non-verbal, or more accurately, non-propositional expressions may likewise be susceptible of classification, and hence permit of counting and statistical analysis. Again, we may classify attitude studies according to the degree of control which the investigator is able to exercise. We thus have at least four types of actual or possible attitude studies which might receive attention in the present survey. Thurstone's studies are of the controlled verbal or propositional type. The studies of "social distance" inaugurated by E. S. Bogardus,[18] and the study by Donald Young [19] of the effects of classroom instruc-

[18] Appearing frequently in *Sociology and Social Research.*

[19] "Some Effects of a Course in American Race Problems on the Race Prejudices of 450 Undergraduates at the University of Penn-

tion in changing student attitudes with respect to race differences, might possibly be cited here as illustrations of controlled but non-propositional studies. Others, more definitely of this type, may easily be conceived. The late lamented silent drama, for instance, presented many situations in which the attitudes of an audience were tested and might have been gauged by such indexes as the sound volume of applause or the ratio of disgusted patrons walking out on the show.

It is in the realm of non-controlled data, whether semi-propositional or otherwise, that the practical need of further and continued attitude studies seems to the writer to be most evident. The difficulties of building scales similar to Thurstone's, and of applying them to the measurement of the attitudes of social groups, become increasingly difficult once we leave the classroom, the discussion club and the other small, comparatively infrequent and highly selected groups that enjoy having experiments tried upon them. Such groups already have developed ways of making their attitudes articulate. It is the more numerous work-a-day groupings of society, which are inaccessible to his controlled measurements, about whose attitudes the social scientist is in the most need of information. Students may be required, good natured academicians may be cajoled, and sundry needy persons may be paid to sort cards containing propositions into eleven piles. But it is difficult to imagine securing comparable judgments, or satisfactory measurements in the final application, from bricklayers, business men, Italian-Americans, nuns, stevedores, or seamstresses. And, unless the scale itself is based upon equal-

sylvania," *Journal of Abnormal and Social Psychology*, XXII, Oct. to Dec., 1927, pp. 235–42.

seeming differences to a random sample of the group which is to be measured, its validity—the degree to which it measures that which it purports to measure—becomes open to question.

There is need, therefore, to examine more carefully the extent to which data established for other purposes, but which to some extent reflect attitudes, may be used as indexes for measurements of the latter. In illustration, such data may be cited as political campaign appeals; election registrations and voting returns; newspaper editorials, news content, advertisements and circulation; purchases of commodities which are dependent upon certain activities, interests, or social habits, such as bibles, bathing suits, hymn books, contraceptives, football tickets, bath tubs, listerine, etc., etc.[20]

Further progress in the statistical attack upon atti-

[20] See Ogburn, W. F., and Talbot, N. S., "A Measurement of the Factors in the Presidential Election of 1928," *Social Forces*, VIII, Dec., 1929, pp. 175–83. Some of the writer's students are engaged in work which is suggestive. A seminar is considering the respective patterns of attitude which are said to characterize and distinguish Philadelphians and New Yorkers. One member has pointed out that New Jersey communities adjacent to New York permit moving picture theaters to open on Sunday, while New Jersey communities adjacent to Philadelphia, subject to the same state laws, require these theaters to close. She is attempting to establish the zone of demarcation between the areas in which the two attitudes here reflected are dominant. Others, by an elaborate experimental classification of radio broadcasts in the two cities, are attempting to determine what types of program in each case are presumably catering to local tastes. Another has suggested comparisons of the library circulations by respective library classification numbers in the two cities and in intervening communities. Another is finding repetitive and classifiable clues to the evaluations placed upon various social attributes by a study of the items of information entered in their autobiographical biographies by state legislators. These represent a few among perhaps two dozen types of uncontrolled and partially non-verbal data which have been discussed as possible indexes of underlying attitudes. Most of these have within them at least the theoretical possibility of statistical analysis.

tudes and opinions depends upon continued development in several directions: (*a*) upon laboratory studies of individual psychology which will aid in further clarification of concepts; (*b*) upon attempts at further empirical verification of the Thurstone scales; and (*c*) upon ingenious statistical analysis of innumerable "uncontrolled" by-products of human activity, which throw light in specific fashion upon the motivations in our minds.

COMMENTARY

L. L. THURSTONE

The galley proof of the preceding article was submitted to Professor Thurstone for comment and criticism. He has been so kind as to prepare a statement which the editor takes much pleasure in appending at this point.

Professor Rice has raised a number of questions that are crucial in the problem of measuring social attitudes. Since several experimental studies have been made with some of these questions particularly in mind, and since these experimental studies have not all been published, a brief statement of the findings might here be in order.

One of the crucial questions that bothered me for some time was to ascertain whether the attitudes of the judges who help to construct the scale affect the scale values of the statements. If so, then the universality of the scale as a measuring instrument would, of course, be very seriously limited. Professor Rice raises this question in several forms. This problem was selected by Mr. E. D. Hinckley for his Doctor's thesis which has not yet been published. He prepared a list of one hundred statements about the Negro, varying from those attitudes which gave him all the social privileges of the white man to those that restricted his social rights. He selected one hundred white college students in Chicago who expressed attitudes favorable to the Negro on five special statements, one hundred white college students in Florida who expressed attitudes restricting the social rights of the Negro, and one hundred educated Negroes. The same set of one hundred statements was sorted by the three groups independently and three scales were constructed. I was very much pleased to see that the three scales so constructed with the same statements were practically identical. This seemed to demonstrate quite effectively that the scale does transcend the attitudes of the judges who sort the statements.

Professor Rice raises the question, "Where is the distinction to be drawn between factual judgment opinions and attitude-representation opinions?" It is, of course, clear that a simple statement of fact which is well known cannot serve as an opinion by which to measure attitude. There are two statistical criteria by which the distinction is rather easily made, namely, the criterion of ambiguity and the criterion of irrelevance. Consider the statement, "Many churches are now being built." What would the judges do with it in sorting? They would have difficulty in sorting it, and its spread or ambiguity would probably be large. That would eliminate it from the scale because it does not convey with sufficient uniformity an attitude favorable or unfavorable to the church. It is a simple statement of fact. Consider the statement, "I don't believe church-going will do anyone any harm." This was statement No. 9 in our first experiment and it was sorted with small ambiguity.

But it was eliminated by the criterion of irrelevance because the pious people as well as those who are antagonistic to the church endorsed it. These two criteria of ambiguity and irrelevance constitute objective methods of drawing the distinction in Professor Rice's question.

The same criteria would eliminate the illustrative statement of Professor Rice about the Negro, namely, "I am impaitent of the Negro's dependence upon sympathetic white men. Force him to work and take care of himself." Such a statement would be interpreted by some judges as reflecting an attitude favorable toward the Negro and by other judges in the opposite manner. Hence, its ambiguity would be large and the statement would not be retained in the scale.

The term attitude has caused some trouble because its meaning has been rather hazy in current writing. I have redefined it recently for the purposes of these experimental studies as *the degree of affect about a psychological object.* Affect has two directions in its primitive sense, appetition and aversion. It is here used to designate potential action which is favorable toward a psychological object or unfavorable and against it. The strength of this affect which may be positive or negative is the trait measured by the attitude scale. It is just as hypothetical as the concept of intelligence which is measured by what it supposedly does. But these concepts are hypothetical in the same sense that the concepts force, momentum and volume are hypothetical in physical science. No one has ever seen or touched a force. It is measured by what it supposedly does. The legitimacy of these abstractions can be tested only in the consistency by which they operate in experience.

That a common core of affect about a psychological object can be postulated which is more or less independent of its detailed rational and verbal expression is indicated by a recent Master's thesis by Eggan. He submitted a list of twenty nationalities for preferential judgment by the method of paired comparison. Six different questions were submitted as, for example, "Which nationality would you prefer to associate with?" and "Which nationality would you rather have your sister marry?" The scale values of the twenty nationalities were practically unaffected by the six different forms of the question. In fact, a subject can proceed to make these preferential judgments about nationalities even after he has forgotten the exact question at the beginning of the schedule. This is as it should be because the preferential judgments are intended to be affective indicators and for this purpose the details of the rational question are really distractors. Affect cannot be conveyed except in some particularized and overt form.

Professor Rice raises the question whether the attitude scales would be affected by the degree of expertness of the group of judges. The statements should be such as occur in the actual population whose attitudes are to be measured and in this sense the scale is limited to one culture. People who cannot read handwriting probably could not judge whether or not it is beautiful. If the statements are long and involved, the scale would not be applicable to those who cannot read complex prose. An experiment should be set up in which a group of experts, a group of educated people, and a group of uneducated people serve as judges in the sorting process. I should venture the prediction that the scales will be identical except that the unit of measurement will be small for the experts and large for the uneducated group. The discriminal error will be smallest for the experts. It is as though the distances between stations along a railroad were expressed in miles and in kilometers. The scales are identical except for the unit of measurement.

It is quite probable that the subject of psychophysics which has hitherto been barren can be vitalized in its application to these psychologically and socially interesting problems. The problems raised by Professor Rice are fundamental. As he has indicated, many of these problems can be solved experimentally so that we may rid the subject of much dialectics.

XII

STATISTICAL STUDIES OF PERSONALITY AND PERSONALITY MALADJUSTMENT

CLIFFORD KIRKPATRICK

THE purpose of this paper is threefold: first, to consider the data respecting personality which are available to the statistical method, and some of the present applications of statistics to their study; second, to consider some of the limitations and advantages of statistical method in relation to the case method; third, to suggest certain possible future applications of the statistical method to the study of personality and personality maladjustment.

I

At the outset, it is necessary to define certain terms and to consider what raw material personality offers for statistical treatment. By the statistical method is understood a technique for collecting, condensing, and deducing the implications of numerical data. The meaning of personality is perhaps not so readily agreed upon. It may be defined for our purposes as the totality of response tendencies manifested by a human organism in various social situations. It should be noted that personality is known to others through behavior, including verbal responses, and through products of behavior. Response tendencies, attitudes, sentiments, complexes, and the like, are merely useful abstractions, based upon observation of behavior, with more or less interpretation in the light of the common conscious experiences of the observers.

It may be tentatively suggested that behavior and the products of behavior, which are all that can be known directly of personality, may be classified under three headings. First, there are *attributes* of personality which may be either present or absent, as for example, a certain political party affiliation, a mental disorder recognized as a definite clinical entity, or citizenship in a certain country. Second, there may be *qualitative variables;* that is to say, things which are present in terms of more or less, but which are not expressible as multiples of a unit which is objective. By objective we mean to designate a stimulus which produces practically the same sense impression upon various observers. Estimates of the degree to which such qualitative variables are present are based upon classification rather than upon counting. Smith is happier than Jones, but not necessarily twice as happy. The sum of just noticeable differences, such as are obtained when a weight is applied to the hand, is not necessarily a multiple of a unitary sensory experience. In general, such personality traits as aggressiveness, introversion, mental energy, optimism, radicalism, and egocentrism are variables which fall into this category.

In the third place, our knowledge of personality may involve *quantitative variables* expressible as multiples of objective units. A man's weight in grams is a multiple of the weight of a cubic centimeter of water at four degrees Centigrade. Measurement of a quantitative variable implies counting units which are actually or conceptually interchangeable. We may count a man's height in inches and assume that an inch at his feet is the same as an inch at his neck. Reaction time in seconds, bricks laid per hour, and standardized problems solved per minute, constitute quantitative variables. The criteria which distinguish a quantitative from a qualitative variable are (1) existence of an objective unit; (2) possibility of counting units to obtain a multiple; and (3) the theoretical interchangeability of units throughout the continuum.

It should be noted that a quantitative variable may be an index of a qualitative variable. We know intelligence as such only in terms of more or less, but the number of problems solved correctly is a quantitative variable which serves as a useful index and one which has important implications. Furthermore, qualitative differences, as between a table and a chair, might be described in terms of a complex pattern of quantitative variables, such as length, breadth, height and proportion ratios. Similarly, differences in musical ability, usually thought of in terms of more or less, might be analyzed in terms of a number of quantitative variables.

It is, of course, possible that the difference between qualitative and quantitative variables is one of degree. In the first place, as Rice has shown, units, even in the physical world, are only relatively identical.[1] Human beings in the United States differ in many respects, yet they are sufficiently similar to give important implications to data based on the assumption that they are identical. In the second place, there may be such objective aids to classification with respect to a qualitative variable that a quantitative variable seems to come into existence.

[1] Rice, Stuart A., *Quantitative Methods in Politics*, New York: Alfred A. Knopf, 1928.

Perhaps the closest approach to such a transformation is represented by a study made by Thurstone in which he claims to have demonstrated that attitudes can be measured. For example, a scale for the measurement of attitudes on pacifism and militarism is devised. It consists of a series of statements assumed to be separated by equal intervals or units of attitude; since judges, on the average, rank them the same number of categories, or the same fraction of a category, apart.[2] Granting that a brilliant contribution has been made to objective methodology, it is not, however, quite certain that a quantitative variable has been evolved. A category, such as sixth in a series of eleven, is not a unit that can be meaningfully multiplied; nor can the unit intervals derived be divided into the behavior of an individual or its products. The opinion of an individual is classified rather than counted; the rating is not a multiple of a unit identical to all individual observers, as is a gram weight; and finally, the difference in opinion between two militaristic statements does not seem conceptually interchangeable with the difference in opinion between two pacifistic statements. While it would seem that the Thurstone scale does not satisfy the criteria of a quantitative variable, the intervals certainly have a conventional meaning similar to that of feet or pounds when used in making mental estimates without the aid of instruments.

It is probable that variables such as score on a spelling test come closer to the borderline between qualitative and quantitative variables, depending

[2] Thurstone, L. L., "Attitudes Can be Measured," *American Journal of Sociology*, XXXIII, No. 4, Jan., 1928, pp. 529–54. Cf. the discussion of Thurstone's methods in the preceding paper herein, on the application of statistical method to the study of social attitudes and public opinion.

chiefly upon the degree to which the words are standardized with respect to difficulty. Nevertheless, there is reason to think that the distinction which has been drawn is at least partially valid and has some measure of utility.

Turning now to the various modes of application of the statistical method to the study of personality, some seven types may be noted.

1. The enumeration of maladjusted personalities and the relating of maladjustment to causative attributes are among the most obvious statistical procedures. Whether the maladjustment is perceived depends, of course, on the social situation. Moreover, many of the traits making for maladjustment are variables. Nevertheless, the delimitation of a variable by a conventional subjective standard is possible, and meaningful statistics have been gathered in regard to the number of feebleminded, insane, criminal, and neurotic persons in the United States. There are fashions in diagnosis, but the fact that certain individuals acted so as to get themselves diagnosed in a particular fashion at a given time has some significance. The survey method of enumeration of the mentally abnormal offers some independence of legal vicissitudes, at least as compared with the investigation of crime.[3] Bailey's study of the feebleminded discharged from the army probably deals with the most representative sample that has been investigated.[4] The various statistical studies of crime are too well known to require further mention.[5]

[3] Rosanoff, A. J., National Committee for Mental Hygiene, Publication No. 9.

[4] Bailey, P., and Haber, R., "Mental Deficiency," Reprint No. 94 from *Mental Hygiene*, IV, No. 3, 1920, pp. 564–96.

[5] Cf. the papers by C. E. Gehlke and L. C. Marshall in the present volume.

2. The study by questionnaire of personality attributes has yielded interesting results, in spite of the limitations and weaknesses of the questionnaire method. The recent work of Katharine B. Davis has given valuable information as to the prevalence of certain sex practices and has related experiences and personality attributes to success in marriage.[6] Wells and Peck have carried out similar studies with college men.[7] The recent study by G. V. Hamilton of the married life of two hundred men and women shows valuable application of the statistical method to questionnaire and case study material.[8]

3. The classification of personalities by qualitative variables represents another application of the statistical method, although it overlaps at times the questionnaire procedure. The classification of personalities by qualitative variables may be with or without objective aids to this ranking process. An example of the second is the rating scheme used in the army and in personnel work. Aggressiveness, ingenuity, reliability, and other traits of an individual, are merely rated by judges on a scale ranging perhaps from one to five. In other cases there are objective aids to the subjective classification of persons with respect to qualitative variables. For example, in Gordon Allport's recent study of political attitudes, various questions were ranked by competent persons as to the degree of radicalism which they indicated. The sub-

[6] Davis, K. B., *Factors in the Sex Life of Twenty-two Hundred Women.* New York: Harper and Bros., 1929.

[7] Wells, F. L., "Report on a Questionnaire Study of Personality Traits," *Mental Hygiene,* IX, No. 1, Jan., 1925, pp. 113–27. Peck, M. W., and Wells, F. L., "Further Studies in the Psychosexuality of College Graduate Men," *Mental Hygiene,* IX, No. 3, July, 1925, pp. 502–20.

[8] Hamilton, G. V., and MacGowan, K., *What is Wrong with Marriage?* New York: Albert & Charles Boni, 1929.

jects were then ranked with the aid of this index, itself arranged by a ranking process.[9] If the application of the criteria of a quantitative variable restricts Thurstone's scale to the qualitative category, the same would apply to Allport's scale, in spite of the numerical value given to degrees of radicalism.

4. The classification of personality with respect to quantitative variables is an application of the statistical method which differs at least in degree. The most spectacular development in this direction is perhaps the mental testing movement which has swept the country. Innumerable tests have been developed to classify the population in regard to general intelligence, general information, mechanical ability, reading ability, and the like. Whether or not there is actual measurement with respect to quantitative variables as a preliminary to classification depends, from the point of view here suggested, upon the degree to which there is an objective unit, upon the possibility of counting such units to obtain a multiple, and upon the theoretical interchangeability of the units. Where answers are given in an intelligence test to questions of equal difficulty, and when the correctness is as demonstrable as the existence of a gram weight, the test score would probably be considered a quantitative variable. When the problems are graduated as to difficulty and the answers are weighted, the score is less definitely quantitative.

Recently there has been a rapid multiplication of so-called personality tests, dealing with emotional traits and social adjustments rather than mere intelligence.[10] Some of these tests make a classification

[9] Allport, Gordon, W., "The Composition of Political Attitudes," *American Journal of Sociology*, XXXV, Sept., 1929, pp. 220–38.

[10] Young, Kimball, "Personality Studies," *American Journal of Sociology*, XXXII, May, 1927, pp. 953–71.

of personalities on the basis of both quantitative and qualitative variables. For example, in the George Washington Social Intelligence Test, the ability to associate names and faces correctly is objectively measured by the number of correct identifications of pictures, presented in an earlier part of the test. In another portion of the test, however, it is required to choose the most tactful solution to the various social situations which are described. Here the rating is essentially on the basis of a qualitative variable, for the subject is classified according to his conformity to solutions which have been subjectively ranked by judges as to their correctness.

5. The establishment of norms is another application of the statistical method—one which may involve classification or other applications previously mentioned. Innumerable illustrations of personality norms come at once to mind. Every mental test, to have meaning, must be standardized to establish norms which may serve as a basis for comparison. The mental age series, the intelligent quotient, and achievement ages in various subjects have been worked out by psychologists and are well known even to laymen. Less familiar are the norms for associative reaction to stimulus words such as those listed in the Kent-Rosanoff table.[11]

6. A slightly different application of the statistical method, although involving the use of norms, consists in the prediction of individual behavior from the norm of the subgroup to which this individual belongs in relation to the norm of the entire group. For example, according to Burgess, a person who has never held a permanent job has less chance for parole success than

[11] Rosanoff, A. J., *Manual of Psychiatry.* New York: John Wiley, 5th edition, 1920, pp. 547–620.

one picked at random.[12] According to Hart's investigation, a person of a higher age group apparently has less chance to avoid divorce than one whose marriage takes place at more nearly the average age.[13] In the same category would be placed the recent study of Ogburn and Winston, on the probability of insanity.[13a]

7. Finally, in regard to the relative influence of heredity and environment in moulding personality, the statistical method has proven itself invaluable. It may even be argued that some form of the statistical method is absolutely essential for even a partial answer to this age-old problem. If the hereditary aspect of personality means anything, it means the potentialities of the original germ cell manifested in organic structure and known through differences and resemblances in personal behavior which exist independently of, or in opposition to, environment. It is futile to debate the relative influence of heredity and environment in producing the neurotic condition of a particular individual. Both are necessary causes. For an evaluation of the rôle of heredity, it is necessary to contrast the reactions of many individuals to the same environmental conditions or to note the resemblances of other individuals in spite of varied environments.

II

We next turn to a discussion of statistical method in relation to case method. There are many limita-

[12] Burgess, Ernest W., "Is Prediction Feasible in Social Work? An Inquiry Based upon a Sociological Study of Parole Records," *Social Forces*, VII, No. 4, June, 1929, pp. 533–45.

[13] Hart, Hornell, and Shields, Wilmer, "Happiness in Relation to Age at Marriage," *Journal of Social Hygiene*, XII, Oct., 1926, pp. 403–7.

[13a] Ogburn, W. F., and Winston, E., "The Frequency and Probability of Insanity," *American Journal of Sociology*, XXXIV, March, 1929, pp. 822–31.

tions of the statistical method which have been pointed out with some vigor by those who have begun to protest against the cramming of every phase of human experience into neat little categories, to be juggled with the aid of mathematical formulae. There has certainly been a tendency to assume a normal distribution where such an assumption is unjustified and ruthlessly to apply formulae based on this assumption, such as that for the Pearsonian coefficient of correlation. There has been, no doubt, a naïve belief in the magic power of mathematics to transform carelessly selected data into eternal verities. A complex pattern of variables is often inadequately analyzed. Control groups are frequently lacking in studies of delinquency or feeble-mindedness. Delinquent children come from broken homes, but in what proportion as compared with children similar in every respect save that they come from stable homes?

The most common criticism of the statistical method as applied to the study of personality, especially from advocates of the case method, is that the selection of a proper statistical unit is either an impossible or a meaningless procedure. This criticism must certainly be given great weight. Consider, for example, the value of the usual statistical material on crime as compared with the wealth of illuminating case material collected by Healy, which gives insight into the details and relationships of human behavior. Personalities described solely in quantitative terms would bear scant resemblance to the vivid personalities presented in their entirety as a complete and dynamic pattern by the case method. The Gestalt psychology but reinforces the ancient truth that the part often depends upon the whole and that an event or personality trait has meaning only with reference to a total situation.

A child goes to the hospital for a tonsilectomy with the complete conviction that she is to be cut open with a butcher knife and killed. How naïve to assume that this tonsilectomy as an experience is identical with the tonsilectomies of other children, and carelessly to group such experiences together as well-defined attributes. Again, a group of persons may be classed together as having the attribute of arrogance. Detailed case studies would doubtless reveal that in certain instances the apparent arrogance is compensatory for inferiority attitudes. Another sample of arrogant persons taken at random might contain a larger proportion with inferiority attitudes, and thus yield blurred or altered correlations even if the distribution of all traits save arrogance remained the same.

Granting various weaknesses of the statistical method in itself, and relative to the case method, four points may, nevertheless, be urged in its favor.

1. The problem of selecting a suitable unit is not an insuperable obstacle. As previously argued, precise definition of statistical units is a matter of degree. Items need be similar merely to the extent warranted by the purpose in mind. Human beings may be treated as identical with respect to seating space in a theater. In regard to diet, however, age, sex, and occupation must be taken into account.

Use of a unit necessarily ill-defined is perhaps not so much an error itself, as a condition which introduces complications regarding sampling, extraneous variables, and uncertain implications; and which paves the way for errors in interpretation of the results. It is interesting to speculate upon the effect of such a unit in a specific instance. Suppose that a sample of arrogant individuals be compared with a control group in regard to percentage convicted of felony, with the result that the relative frequency is twice as great for the arrogant group. In another less representative sample there might be a slightly different proportion showing purely compensatory criminal behavior. But given the same method of selection in a large number of cases, the extraneous variables would tend to cancel each other and the proportion of arrogant would tend to become stabilized. The implications of such a hypothetical study as to the behavior of other groups and the probable behavior of individuals would largely depend, then, upon representative sampling. The degree to which sampling is representative must be considered in interpreting any statistics, and such sampling is merely made more difficult by the relatively uncertain units descriptive of personality traits. The case method could separate such attributes as real and compensatory arrogance and render aid in making sampling more representative and implications more precise.

2. The statistical method has an advantage over the case method in regard to the tracing of causal relations between events and personality traits. Causation implies persistent association, and to detect such a relationship there must be repetitions. In so far as repetitions are noted in a case study, there is used what might be called an *informal statistical method*.[14] The student observes a series of behavior sequences and makes an informal deduction of a causal relationship operating in the particular case, or else applies the experience derived from a knowledge of other cases, which is again an informal statistical procedure.

[14] My attention has been called to the fact that Bernard has previously used this term. See: Bernard, L. L., "The Development of Methods in Sociology," *The Monist*, Reprint, April, 1928, p. 310.

3. While it may be granted that the case method sets problems for statistical treatment, the statistical method in turn paves the way for an intensive study of particular cases. The statistical fact that a greater proportion of mental disorders is found among the unmarried suggests the use of the case method in seeking clues for further statistical analysis. The informal statistical method is used constantly in planning an approach to the study of a particular case. Who would think of looking for an inferiority complex behind assertive behavior, unless such a paradoxical coexistence had been noted in other cases?

4. The outstanding advantage of the statistical method is the fact that it overcomes the fallacy of generalization from insufficient cases, which is the fundamental limitation of the case method. Again and again, we find vast deductions made from three or four clinical cases. One has the feeling that there are other cases in the background of the writer's experience, but these are not formally recorded and do not constitute an orderly marshalling of objective evidence to prove the alleged relationship. The study of a particular case has implications only for that case, and, assuming determinism, for identical cases. There is intuitive knowledge that many other individuals must fall roughly into the same category and be subject to the implications of the single case. The statistical method merely renders formal and precise the processes necessary either for the meaningful study of the individual case or for the projection of conclusions to other cases which are similar in some respect.

III

Our consideration of probable future applications of statistical method to the study of personality and personality maladjustment follows logically from the preceding discussion. There is an enormous amount of case material in the files of various organizations from which only informal deductions are made and which awaits statistical analysis. More could be gathered by the use of suitable schedules. By accepting the uniform judgments of competent social workers, reasonably homogeneous attributes could be isolated. There is reason to think that, pending the further development of objective personality tests, far more extensive use may be made of the method of classification by qualitative variables, using the technique of rank order classification. If an individual is to be described in a case record as egocentric, why should he not be rated as to the degree of his egocentrism in relation to other cases known to the worker? To be ranked four on a scale of five is more significant than to be merely labeled egocentric. Frank has recently made an effective plea for a further application of the rank order procedure. While a psychogram indicating the rank or score of an individual in regard to various personality traits would fail to picture the entire personality as a functional unity, it would nevertheless have many advantages when used in conjunction with the conventional case method.[15]

The limitations of the case method are clearly shown by the claims of the so-called newer psychology. Many of these are by no means convincing, due to the lack of statistical procedure except of the most informal character. With the aid of objective tests, ratings, and checks upon subjective definitions of units similar to the coefficient of aggregate variation

[15] Frank, L. K., "Personality and Rank Order," *American Journal of Sociology*, XXXV, Sept., 1929, pp. 177–86.

recently developed by Rice, much may be accomplished in the study of personality by the statistical method.[16] Many of the ingenious speculations of the various brands of psychoanalysis as to the processes moulding personality and determining personality maladjustment may be refuted or verified. Only a comparatively few of the various fields of statistical investigation that may be fruitfully cultivated in the future, may be indicated here. Some ten possibilities will be mentioned, some of which have already been partially realized, perhaps to a greater extent than is known to the writer.

1. The effects of infantile experiences upon later personality have figured largely in various systems of psychological theory. Case records frequently contain data in regard to such matters as premature birth, bottle feeding, and use of instruments in delivery; as also to early diseases, malnutrition, physical injury, etc. Back of the record is the apparent belief that these conditions are related in a significant manner to subsequent personality. This is indeed possible, but for proof it would be necessary to make statistical investigations, probably with the use of a method of paired comparisons. With the aid of case records, individuals should be paired who are similar in as many respects as possible, except that they differ in regard to the specific infantile conditions which are to be investigated.

2. The reference to physical injury brings to mind the possibility of checking in a scientific manner certain of the basic assumptions in Alfred Adler's psychology. Significant physical defects, that is to

[16] Rice, Stuart A., and Weaver, Wallace, "The Verification of Social Measurements Involving Subjective Classifications," *Social Forces*, VIII, No. 1, Sept., 1929, pp. 554–58.

say, those so regarded by competent observers with respect to the entire social situation, could be checked as to their effect upon adult personality, compensatory trends, and the like. Here again the method of paired comparisons could be utilized to good advantage to control other variables. Even the questionnaire method would throw some light upon the operation of the so-called masculine protest among females as compared with males. A recent study of Aikins, using for the most part the questionnaire method, has shown the prevalence of the so-called inferiority complex among women; but there is no check as to the relative frequency among men, and as to the degree to which the inferiority feelings may be attributed to female status.[17]

3. There is every reason to acknowledge the profound influence of abnormal family situations upon the personality of the growing child. There is, however, a lack of precision in most of the statistical studies thus far made, while the case studies are supported merely by the informal statistical method. Numerous fascinating problems present themselves as to the effects upon the child of the widowhood of the mother, of only-child status, of lack of security, of preferences by parents, of family tension, of low relative family status, of extreme domination by the father, and of jealousy-producing situations. Certain of these situations would constitute fairly well-defined attributes; and extraneous factors could be controlled in part, again by the method of paired comparisons. Certain of the family situations might be described in terms of quantitative variables, such as the number of years of close contact with one parent, the age to

[17] Aikins, H. A., "Women and the Masculine Protest," *Journal of Abnormal and Social Psychology*, XXII, No. 3, Oct. to Dec., 1927, pp. 259–72.

which the child slept with his mother, the approximate frequency of caresses, the frequency of threats of a certain type, or the frequency of punishments, precisely defined. The statistical method would, of course, be reinforced by questionnaires and case material. Certain phases of adult personality presumably affected by these conditions might be objectively, even quantitatively, described, with the aid of rating scales, personality tests, and classification by competent observers, as well as by questionnaire methods. The number of years of postponement of marriage, in the case of sons of widowed mothers, would be a clear-cut quantitative variable.

4. Aside from physical defects and family situations, there are psychic experiences of traumatic character revealed in case records which might well be correlated with subsequent behavior by statistical methods. We know that certain children develop phobias after having been severely frightened by animals. What proportion of children suffering such an experience between certain ages develop a phobia, as compared with a paired group who have not suffered any such experience? Intense ego injuries or humiliating experiences, as recognized by competent observers in the light of the entire background, might also be correlated with later personality trends.

5. In view of Freudian doctrines, the effect of early sex experiences is of extreme importance. Healy has shown by case methods that associations connected with early sex experience probably explain subsequent delinquencies, although for the individual case we have little more than a *post hoc ergo propter hoc* relationship.[18] Does a group of individuals having

[18] Healy, W., *The Individual Delinquent.* Boston: Little, Brown, 1917.

early sex experiences of approximately the same type show differences of personality traits in kind or degree from a corresponding group having early sex experiences of an entirely different character? It would be interesting to know more as to the frequency with which knowledge of the sex relations between parents comes to children as a severe shock, productive of mental conflict. The questionnaire and case methods might furnish material for statistical treatment.

6. The newer psychology attaches much importance to dream and revery on the part of both children and adults. Illuminating case studies have been presented bearing on this point. It might be possible, however, to classify dreams and reveries into certain types and refer them to social situations and to various attributes of personality. Beginnings have been made in this direction which illustrate the possibilities. Conklin has shown the foster-child fantasy to be remarkably prevalent in the childhood experience of college students.[19]

Kate Gordon has shown that the dreams of orphan children tend to deal with parents and with home life.[20] The proportion of dreams or "slips" of a certain type, and the frequency of certain mannerisms, might be related to personality traits. Furthermore, as Goodenough has shown, children's wishes expressed in a spirit of play have considerable diagnostic significance and might well receive statistical treatment.[21]

7. There might well be further investigation of personality types with the aid of statistical methods.

[19] Conklin, E. S., *The Foster-Child Fantasy*. Eugene: University of Oregon Publications, Vol. I, No. 4, 1920.

[20] Gordon, Kate, "Dreams of Orphan Children," *Journal of Delinquency*, VIII, Sept. to Nov., 1923, pp. 287–91.

[21] Goodenough, Florence, "The Diagnostic Significance of Children's Wishes," *Mental Hygiene*, IX, No. 2, April, 1925, pp. 340–45.

Spearman has used certain correlation methods in his study of an alleged general factor in intelligence.[22] Gordon Allport, in the study previously mentioned, has shown that political attitudes are by no means separate and specific entities, and his work suggests the possibility that correlation studies may show fairly well-defined personality types.

8. Statistical investigation of cycles of personality change in relation to social conditions and physiological rhythms is a fruitful possibility. An elation-depression scale upon which to rate the individual's mood at appropriate time intervals might bring to light interesting correlations with sex habits and ego experiences, as well as with more distinctly physiological conditions. Hersey has undertaken a pioneer study in this direction.[23]

9. Few aspects of human behavior are more perplexing than the behavior of falling in love. The Freudian concepts are unquestionably the most useful explanatory tools that have thus far been forged. Here again statistical verification or refutation is necessary. It seems quite feasible to make studies, at least partly statistical, of the relationship between the personality of the parent of opposite sex and the personality of the actual or desired mate. This could be done quite readily for physical traits, and quite possibly for other traits, with the aid of personality tests and rating scales. Steps in this direction have been made by Hamilton and MacGowan in their recent study.[24]

[22] Spearman, C. E., *The Abilities of Man, Their Nature and Measurement.* New York: *The Macmillan Co.*, 1927.

[23] Hersey, Rex B., "Periodic Emotional Changes in Male Workers," *Personnel Journal*, VII, No. 6, April, 1929, pp. 459–64; also, "Cycles in Workers' Efforts and Emotions," *Engineers and Engineering*, XLVI, No. 7, July, 1929, pp. 162–66.

[24] *Op. cit.*

10. Finally, the statistical method may be usefully applied to the correlation of personality traits with success in marriage, as rated by a suitable scale. As previously mentioned, Hart has dealt with a simple factor, the age of marriage, and others have been considered by Davis in her questionnaire study. To what degree are similarities between mates with respect to self-assertion, intelligence, introversion, temper, integration, and social background related to their success in marriage? These are only a few of the problems that cry out for solution with the aid of the statistical method.

In conclusion, it should be noted that the case method and the statistical method must be regarded not only as supplementary but as complementary. The case method aids in defining units and in arranging paired comparisons; it provides raw material and gives clues to statistical investigations. The statistical method, on the other hand, checks hypotheses formulated from case material, aids informally in tracing behavior sequences in the case of an individual, traces causal relationships, expands the implications of case studies, and facilitates comparison and classification of personalities. To say that the statistical and the case methods are two aspects of the same process is to deviate but little from the truth.

INDEX

INDEX

INDEX

www.ingramcontent.com/pod-product-compliance
Lightning Source LLC
LaVergne TN
LVHW011819200926
848484LV00005B/1244